Construction Of Hot Mix Asphalt Pavements

Manual Series No. 22
Second Edition

ASPHALT INSTITUTE

HMA CONSTRUCTION

Useful Publications of the Asphalt Institute

SP-1   *Performance Graded Asphalt Binder Specification And Testing*

SP-2   *Superpave Mix Design*

MS-1   *Thickness Design – Asphalt Pavements for Highways And Streets*

MS-2   *Mix Design Methods for Asphalt Concrete And Other Hot Mix Types*

MS-4   *The Asphalt Handbook*

MS-5   *Introduction to Asphalt*

MS-6   *Asphalt Pocketbook of Useful Information*

MS-10   *Soils Manual*

MS-11   *Thickness Design – Full-Depth® Asphalt Pavements For Air-Carrier Airports*

MS-12   *Asphalt in Hydraulics*

MS-14   *Asphalt Cold Mix Manual*

MS-16   *Asphalt in Pavement Maintenance*

MS-17   *Asphalt Overlays for Highway and Street Rehabilitation*

MS-18   *Sampling Asphalt Products for Specifications Compliance*

MS-19   *Basic Asphalt Emulsion Manual*

MS-21   *Asphalt Cold-Mix Recycling*

MS-23   *Thickness Design – Asphalt Pavements For Heavy Wheel Loads*

For complete listings of Asphalt Institute publications, videos and slide shows, visit our website at asphaltinstitute.org, or ask for our current catalog by writing to: Asphalt Institute, Research Park Drive, P.O. Box 14052, Lexington, KY 40512-4052, USA.

Printed in the USA

This manual has been prepared to provide the essential information required for the quality control of hot mix asphalt (HMA) pavements. Emphasis has been placed on information which will benefit asphalt construction inspectors, laboratory technicians, plant personnel and paving crews. Since the manual includes technical information on all aspects of HMA pavement construction, it is an excellent reference for anyone involved in asphalt paving operations.

This second edition has been largely rewritten. Chapter 1 has been rewritten as a different, but related, subject from the original version. Chapters 7 and 8 on "Quality Control and Acceptance of Hot Mix Asphalt" and "Segregation," respectively, have been added. The writing and editing of this manual has been a cooperative effort of the Asphalt Institute field engineers, along with the help and cooperation of the headquarters.

The standard practice of the Asphalt Institute is to utilize the International System of Units (SI) (metric), followed by U.S. customary units.

The Asphalt Institute does not endorse products or manufacturers. Trade or manufacturers' names appear herein only where they give credit for photos or other information and where it is necessary for the purpose of this publication.

ASPHALT INSTITUTE
Research Park Drive
P.O. Box 14052
Lexington, Kentucky 40512-4052
(606) 288-4960
Fax (606) 288-4999

FOREWORD

iii

Chapter 1: Construction Project Management 1-1

Chapter 2: Materials 2-1

Chapter 3: Mix Design 3-1

Chapter 4: Plant Operations 4-1

CONTENTS

Chapter 5: Placing Hot Mix Asphalt 5-1

Chapter 6: Compaction 6-1

Chapter 7: Quality Control and Acceptance of Hot Mix Asphalt 7-1

List of Figures

List of Tables

Introduction

Each year, billions of dollars are spent on hot mix asphalt (HMA) construction projects in the United States. Achieving good performance of these pavements does not just happen, but is the result of many hours of effort and project management by pavement design and construction professionals. Each phase of a project, from the drawing board to the laboratory to the field, contains important steps and procedures that must be followed to ensure a long lasting, well performing pavement.

Project management has evolved as a tool to plan, coordinate and control the complex and diverse activities of current HMA construction projects. Factors leading to new project management techniques include:

- Economic pressures within the transportation industry.
- Competition between rival contractors.
- A much higher regard for the value and well-being of the workforce and environment.
- Implementation of stringent quality control standards.
- Safety for the construction workforce and public transportation users.

As the title indicates, this manual is devoted to the principles of "Construction of Hot Mix Asphalt Pavements." It is the intent of the manual to provide the technical information necessary to carry out sound construction project management for HMA pavements. The manual has been written for two distinct, yet similar groups of individuals: those employed by the owner or agency of the construction project, and those employed by the construction contractor. Not all personnel from both groups need to know all the information contained in this manual. Obviously, it is necessary to be familiar with the portion directly related to an individual's job. However, an understanding of the entire design and construction process will often provide an appreciation for where a specific task fits into the overall project management process, and thereby ultimately improve in-service pavement performance.

Purpose of Construction Project Management

The primary purpose of HMA construction project management is to foresee and predict as many of the dangers and problems as possible and to plan and control activities so the project may be completed successfully in spite of all the risks. This process starts before any resources are committed, and must continue until all work is finished. The objective is for the final result to satisfy the project owner within the promised time period and without using more dollars and other resources than those originally established by a contract agreement.

Planning and control must be exercised over all the activities and resources involved in the HMA construction project. This can only be accomplished through effective communication. The communication process requires an understanding of the project documents by both the owner representatives and the contractor personnel. It embodies a whole framework of logical and progressive planning and decisions, perceptiveness, a liberal application of common sense, proper organization, painstaking attention to documentation, and a clear grasp of proven and long-established principles of construction project management.

HMA construction projects have in common the fact that the accomplishment phase must be conducted on a site exposed to the elements and remote from the contractor's facility. These construction projects incur special risks and problems of organization and communication. They often require massive capital investment and deserve (but do not always get) rigorous management of progress, finance and quality.

Construction Project Management is carried out to ensure quality work in compliance with the project requirements. This entails employing active communication in organizing, planning and executing HMA construction projects in accordance with the written instructions detailed in the formal project documents.

First and foremost, both agency and contractor personnel must address the importance of communication concerning the project contract. The *contract* is the agreement between the agency and the contractor. It states the obligations of both parties including criteria for labor, materials, performance and payment. While there are many documents that make up the HMA construction contract, the agency's project manager is concerned primarily with the plans and specifications. Together, the plans and specifications explain requirements that the contractor must fulfill to construct a satisfactory product and be paid in full for the work. Special provisions within the contract normally relate to addenda or supplemental specifications that are project specific.

Plans are the contract documents that show the location, physical aspects and dimensions of the work to be accomplished. They include layouts, profiles, cross-sections and other details.

Specifications are the written technical directions and requirements for the project. *Standard specifications* are included to complement the plans by providing instructions that are not indicated on the drawings. The items normally included in the standard specifications describe the method and manner of executing the work. These specifications normally describe quantities and the quality of materials and labor to be provided under the contract. Material specifications and test procedures from the American Society for Testing and Materials (ASTM) and the American Association of State Highway and Transportation Officials (AASHTO) are listed in the standard specifications and are legally part of the contract documents by reference only. Other documents, such as the Manual on Uniform Traffic Control Devices (MUTCD) and the Occupational Safety and Health Administration (OSHA) regulations are also included in the standard specifications and become a legal part of the contract documents by reference only. Specifications are the means of communication among the contractor and the construction project manager. Specifications are particularly important to the agency and contractor representatives, as they constitute the rules by which the HMA project is managed and accomplished.

Agency/Contractor Relationship

The 21st Century ushers in a new relationship with the agency/contractor in the HMA construction industry. The new relationship is based upon an increasing recognition of the importance of quality during the 1990's. As a result, the Total Quality Management (TQM) concept is being adopted as the key to the HMA industry's ability to compete in today's environment. Partnering, training programs, certification requirements, etc. are integral parts of the overall TQM concept, which involves the following key elements:

- Top management must be committed to the TQM process.
- Quality is derived from the identification, improvement, and control of those processes essential to HMA construction.
- Statistical analysis is necessary to achieve TQM.
- Teamwork and cooperation of all involved is necessary to produce total quality.
- Continual improvement in quality is the focus for TQM.
- Agency/owner satisfaction is the ultimate goal.

In the past, many state and other agencies have performed all of the sampling, testing and approval of materials and processes that were incorporated into the project. In the current environment, many agencies utilize Quality Assurance (QA) type specifications. Under this type of specification, the contractor furnishes the HMA mix designs performed either by its technicians or those of a qualified firm. The contractor is further required to accomplish the quality control (QC) testing on the HMA production and densification. The agency will have a program of acceptance sampling and testing normally performed either by its personnel or a qualified testing firm. In some acceptance programs, the agency will use a combination of contractor test results along with verification tests. The acceptance program confirms the test results of the contractor and gives the agency the documentation necessary for final acceptance and payment for the product. The use of QA/QC specifications has resulted in an improved HMA product. Maintaining and improving the quality of construction will achieve a substantial increase in the pavement life for future HMA construction.

In addition to contractor QC activities, practically all asphalt binder suppliers and many aggregate suppliers conduct extensive testing of their products prior to shipping them to HMA facilities. Virtually all asphalt binders are shipped under certification programs controlled by state agencies. Various programs are employed by aggregate suppliers in cooperation with state agencies to control and verify the quality and gradation of aggregates used in HMA production.

TQM for HMA Construction

The two key TQM processes for HMA construction are hot mix plant production and on-site HMA placement. It is important to the control of a process to establish those characteristics that will lead to a quality product. It would be desirable if a single important characteristic could be chosen for a selected process; however, in the case of HMA construction the processes are too complicated to be properly controlled with just one characteristic. The difficulty of controlling a process drastically increases with the number of measurement characteristics. Therefore, it becomes necessary to minimize the number of chosen characteristics. In addition, there should be precise test methods for each measurement characteristic. The selection of processes, test characteristics and test methods are basic to the quality program, which is only as good as their selection.

In the hot mix plant production process, the aggregate is graded to specific sizes, the mix is heated to a specified temperature, and the aggregate and asphalt binder are mixed to adequately coat each aggregate particle. Past experience has demonstrated that gradation of the aggregate and control of the asphalt binder content are the most important requirements for the control of HMA plants. The ignition test (which is replacing the extraction test) has proven valuable to check the asphalt binder content. Proper calibration either by weight or volume is relatively easy with modern HMA plants to assure the accuracy of both aggregate gradation and asphalt binder content. The accuracy of the measuring devices of a hot mix plant should be checked before the plant is permitted to start producing HMA for a new project. Most modern hot mix plants have computerized controls to permit a running record of the asphalt binder content of the HMA.

The two most important characteristics in the HMA placement process are smoothness and air voids (related to density). Smoothness of the pavement is the most important characteristic for the driving public or agency satisfaction. The driving public is happy as long as the pavement remains smooth and has good traction. Continuous measurement of smoothness is an essential quality requirement.

Proper compaction is paramount to the durability and smoothness of the pavement, and to the achievement of the proper air void content. The volumetric properties of the compacted HMA mixture are the key elements that control the final performance of the HMA pavement. Frequent determinations of the volumetric properties are mandatory to control the placement and compaction process.

Thus far very little has been mentioned about aggregate gradation control. The best way to assure aggregate gradation control is to carefully calibrate the HMA plant. It must be understood that aggregate gradation is closely related to the volumetric properties of the compacted mix. Provided the asphalt content is closely controlled and the HMA is properly compacted, the volumetric properties of the compacted mix provide a measure of gradation variation. In the cases where the contractor's compaction operation is achieving uniform volumetric properties in the pavement, these properties can be used as a measure of aggregate gradation control.

The use of up-to-date techniques enhances the ability to produce quality HMA. Control charts of the asphalt content, aggregate gradation, smoothness, and selected volumetric properties should be maintained and studied to control and improve plant operation. When a control chart indicates problems, immediate attention should be initiated to achieve improved quality. In addition, many of the previously laborious records and reports can be avoided by the routine use of the computer and computerized data recording.

Sampling and Testing

Sampling and testing are methods of evaluating the quality of work. The agency and contractor must agree at the pre-construction conference what sampling is to be performed at the plant and at the roadway, the manner and location in which samples are to be taken, and the number of samples required for a given unit of work. It is the responsibility of both parties to ensure that representative samples are obtained. The samples must be properly identified with the time and date and the location of the source. Both parties must be familiar with quality assurance procedures for sampling and testing. These procedures must be followed to ensure accurate results. If laboratory testing of samples is required, follow-up action should be taken to ensure that tests are performed as scheduled and that the results are properly evaluated.

Safety

Safety is the business of everyone on the project. The project manager must be alert to ensure that safe working conditions and practices are maintained on the project at all times. Safety begins with the project manager, who should set an example in the use of personal safety equipment such as hard hats, gloves, eye protection and protective clothing. In addition, the project manager must see that the safety requirements of the contract are adhered to. This may involve monitoring equipment operation and using such items as barricades, warning lights and reflectors.

Conclusion

Total quality is the key to better performing pavements and satisfying the driving public. With heavier wheel loads and greatly increased average daily traffic expected in the future, TQM is the key to success. Had industry not made the previously mentioned improvements to meet future demands, highway deterioration would be significantly greater than it presently is. History has shown that no matter what level of funding the infrastructure receives, it remains underfunded. Agencies must do more with less for the 21st Century.

In conclusion, this translates into both the agency and the contractor having to work smarter. The material suppliers also play a major role in that they must consistently supply a better product than in the past. The era of Superior Performing Asphalt Pavements (Superpave) is with us. The wheels of progress and technology must continue to turn. The many men and women who perform their duties at the project level will find this manual useful in the discharge of their daily duties.

The modern use of asphalt for road and street construction began in the late 1800s and grew rapidly with the emerging automobile industry. Since then, asphalt technology has made giant strides. Today, highly sophisticated equipment and techniques are used to build hot mix asphalt (HMA) pavements. One rule that has remained constant throughout asphalt's long history in construction is this: *A pavement is only as good as the materials and workmanship that go into it.* No amount of sophisticated equipment can make up for use of poor materials or poor construction practices. This chapter is a discussion of materials used in quality HMA pavements – what they are, how they behave and how to tell whether or not particular materials are suitable for a paving project. The objectives of this chapter are to:

- Understand the properties of asphalt cement and the asphalt grading systems used.
- Recognize the principal tests for identifying certain properties of asphalt.
- Know the procedures for safe and proper storage, handling and sampling of asphalt.
- Recognize the proper methods for handling and stockpiling aggregates.
- Understand the various aggregate classifications and sources.
- Understand certain aggregate properties and the evaluation of aggregate test procedures.

General Information

Asphalt pavements are composed of two materials: asphalt and aggregate (rock). There are many different types of asphalts and aggregates. Consequently, it is possible to make different kinds of asphalt pavements. Among the most common types of asphalt pavements are:

- Dense-graded hot mix asphalt
- Open-graded surface and base courses
- Stone-filled mixes
- Sand hot mix asphalt
- Sheet hot mix asphalt
- Asphalt emulsion mixes (cold mixes)

Hot mix asphalt pavement is the highest quality among the different types. It consists of well-graded aggregate and asphalt cement, which are heated and blended together in exact proportions at a hot mix plant. When all the aggregate particles are uniformly coated, the HMA is hauled to the construction site where an asphalt paver places it onto the prepared roadbed. Before the mixture cools, rollers compact (densify) it into a final pavement to achieve a specified density.

Other pavement types are produced and placed in similar ways. Cold mix pavements use asphalt emulsion or cutback asphalt in accordance with local environmental guidelines. They require little or no heating of materials and can often be produced at the construction site without a central plant. Only HMA is discussed in this manual.

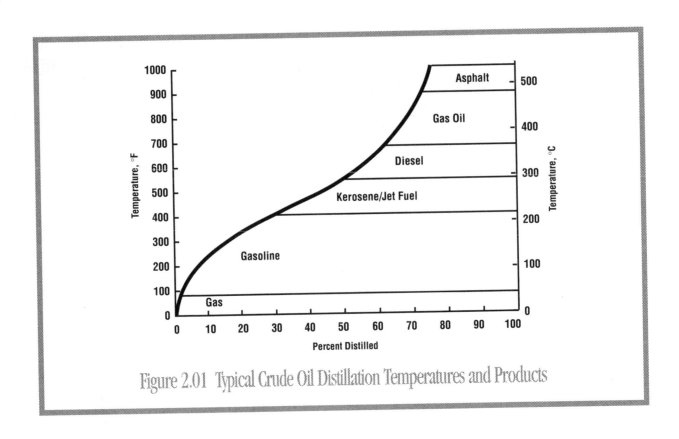

Figure 2.01 Typical Crude Oil Distillation Temperatures and Products

Asphalt

Asphalt is a black, cementitious material that varies widely in consistency from solid to semisolid (soft solid) at normal air temperatures. Asphalt is made up largely of a hydrocarbon called bitumen. Consequently, it is often called a bituminous material. Asphalt is also a thermoplastic material because its consistency changes as its temperature changes. When heated sufficiently it softens and becomes a liquid, thus allowing it to coat aggregate particles during hot mix production. When it cools, asphalt hardens and holds the particles together.

At one time natural asphalt was available, but virtually all asphalt used today is produced by petroleum refineries. The degree of control allowed by modern refinery equipment permits asphalt production for specific applications. As a result, different asphalts are produced for paving, roofing and other special applications.

Paving asphalt, commonly called asphalt cement or asphalt binder, is a highly viscous (thick), sticky material. It adheres readily to aggregate particles, making it an excellent cement. Asphalt cement is an excellent waterproofing material and is unaffected by most acids, alkalis (bases) and salts. This means that a properly constructed HMA pavement is waterproof and resistant to many types of chemical damage.

►►Source and Nature of Asphalt

Because asphalt is available in many types for various purposes, there is sometimes confusion about where asphalt comes from, how it is produced, and how it is classified into grades. The purpose of this section is to describe the source and nature of asphalt types suitable for hot mixing with aggregate for pavement construction. A glossary of common terms related to various asphalt types is found in Appendix B of this manual.

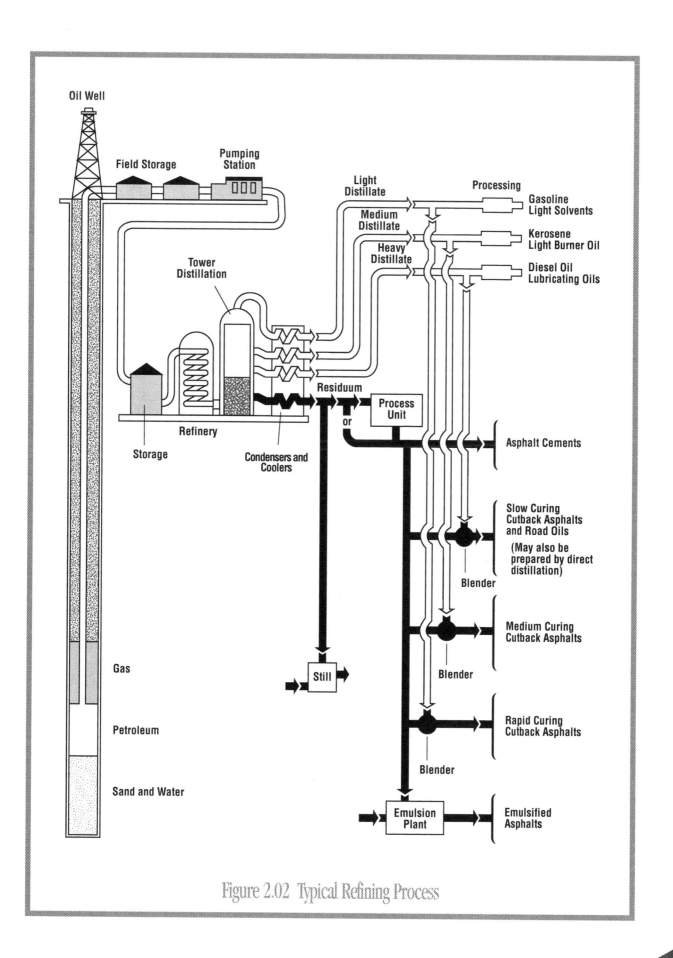

Figure 2.02 Typical Refining Process

Petroleum Refining The production of asphalt begins with the petroleum refining of crude oil. Crude oil is processed by a refinery and is initially separated into several fractions or parts by high temperature distillation as shown in Figure 2.01. Heavy fractions are produced into asphalt while lighter fractions are processed into gasoline, kerosene, diesel fuel, jet fuel and other petroleum products. First, the crude oil is heated to approximately 700°F before being released as mostly vapor into a fractionation tower. The tower contains several temperature zones to permit the vapor to condense into various fractions. A second heater and tower at 700°F may also be employed to simulate much higher temperatures by drawing a vacuum on the tower. This results in further separation into additional fractions. Each fraction is directed through further processing to make modern day fuels and other petroleum products. Figure 2.02 is a simplified illustration of a refinery showing the flow of petroleum during the refining process.

Asphalt Refining Different types and grades of asphalt are required for various applications. To produce asphalts that meet specific requirements, a refinery employs one or more methods to achieve the properties of the desired grade. This is often accomplished by blending crude oils from various sources together before processing (when these crude oils are known to yield the desired properties).

Further control of asphalt properties can be achieved when the crude oil is processed by high temperature and vacuum distillation. Subtle adjustment to temperature and vacuum level affects the amount of soft compounds removed such that the heaviest fraction will exhibit the desired physical properties of asphalt. Some refineries are equipped with solvent-based extraction units, which are effective in removing the soft compounds from the heavy distillation fraction. Solvent extraction of soft compounds usually yields an asphalt product which is very hard and stiff – as desired for certain industrial applications. This hard asphalt can also be blended with softer or less stiff asphalt produced from other crude oils, or at other refineries, to make the final desired asphalt grade.

In summary, crude oil or crude oil blends are used to produce asphalt with specific characteristics. Asphalt is separated from the other fractions by vacuum distillation or solvent extraction and then blended with other asphalts as needed to meet the desired specification.

➤➤Asphalt Behavior

Asphalt is a *viscoelastic* material. This means that asphalt has the properties of both a viscous material and an elastic material. The property that asphalt exhibits, whether viscous, elastic or most often a combination of both, depends on *temperature* and *time of loading* (Figure 2.03). The flow behavior of an asphalt could be the same for one hour at 60°C (140°F) or 10 hours at 25°C (77°F). In other words, the effects of time and temperature are related. The behavior at high temperatures over short time periods is equivalent to what occurs at lower temperatures and longer times.

High Temperature Behavior In hot conditions (e.g., desert climate) and/or under sustained loads (e.g., slow moving trucks), asphalt cements behave and flow like *viscous* liquids (Figure 2.04). Viscosity is the material characteristic used to describe the resistance of liquids to flow. Viscous liquids, like hot asphalt, are sometimes called *plastic* because once they start flowing they do not return to their original position.

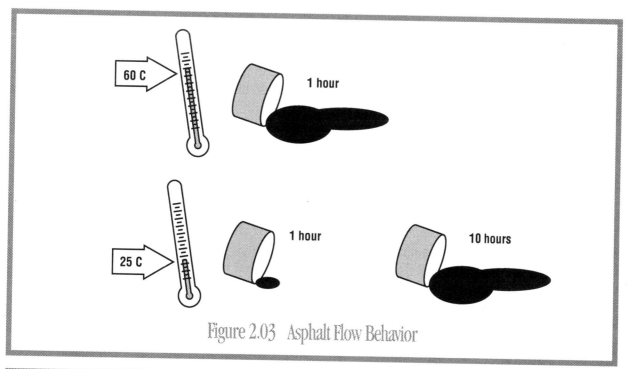

Figure 2.03 Asphalt Flow Behavior

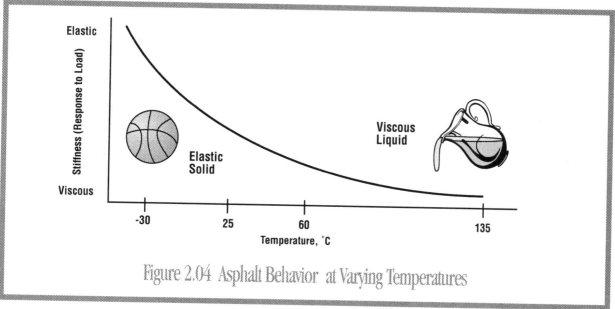

Figure 2.04 Asphalt Behavior at Varying Temperatures

Low Temperature Behavior In cold climates or under rapid loading (e.g., fast moving trucks), asphalt cements behave like *elastic* solids (Figure 2.04). Elastic solids are like rubber bands. When loaded they deform and when unloaded they return to their original shape. Any elastic deformation is completely recovered.

Low temperature cracking sometimes occurs in asphalt pavements during cold weather. In these cases non-load related internal stresses accumulate in the pavement as it tries to shrink and is restrained (e.g., as when temperatures fall during and after a sudden cold front).

Intermediate Temperature Behavior Most environmental conditions lie between the extreme hot and cold situations. In these climates, asphalt cements exhibit the characteristics of both viscous liquids and elastic solids. Because of this range of behavior, asphalt is an excellent adhesive material for paving, yet an extremely complicated material to understand and explain. When heated, asphalt acts as a lubricant, allowing the aggregate to be mixed, coated and tightly-compacted to form a smooth, dense surface. After cooling, the asphalt acts as the glue to hold the aggregate together in a solid matrix. In this finished state, asphalt is considered viscoelastic. It has both elastic and viscous characteristics, depending on the temperature and loading rate.

This kind of response to load can be related conceptually to an automobile's shock absorbing system. These systems contain a spring and a liquid filled cylinder. The spring is elastic and returns the car to the original position after hitting a bump. The viscous liquid within the cylinder dampens the force of the spring and its reaction to the bump. Any force exerted on the car causes a parallel reaction in both the spring and the cylinder. In hot mix asphalt, the spring represents the immediate elastic response of both the asphalt and the aggregate. The cylinder symbolizes the slower, viscous reaction of the asphalt, particularly in warmer temperatures. Most of the response is elastic or viscoelastic (recoverable with time), while some of the response is plastic and non-recoverable.

Aging Behavior Because asphalt cements are composed of organic molecules, they react with oxygen from the environment. This reaction is called oxidation and it changes the structure and composition of asphalt molecules. Oxidation causes asphalt cement to harden, hence the term oxidative hardening or age hardening.

In practice, some oxidative hardening occurs before the asphalt is placed. At the hot mix facility, asphalt cement is added to hot aggregate and the mixture is maintained at elevated temperatures for a period of time. Because asphalt cement exists in thin films covering the aggregate, the oxidation reaction occurs at a faster rate. "Short term aging" is used to describe the aging that occurs in this stage of the asphalt's life.

Oxidative hardening also occurs during the life of the pavement because of exposure to air and water. "Long-term aging" happens at a relatively slow rate in a pavement, although it occurs faster in warmer climates and during warmer seasons. Because of this hardening, old asphalt pavements are more susceptible to cracking. Improperly compacted asphalt pavements may exhibit premature oxidative hardening because of a higher percentage of interconnected air voids. This allows more air to penetrate into the asphalt mixture, increasing oxidative hardening.

➤➤ Asphalt Classification and Grading

Classification of Asphalt Paving asphalts are classified into three general types:

- Asphalt cement
- Asphalt emulsion
- Cutback asphalt

Each is defined in Appendix B. Cutbacks and emulsions are used almost entirely for cold mixing and spraying and will not be discussed further in this chapter.

Because of its chemical complexities, asphalt specifications have been developed around physical property tests, such as penetration, viscosity and ductility. These tests are performed at standard test temperatures, and the results are used to determine if the material meets the specification criteria.

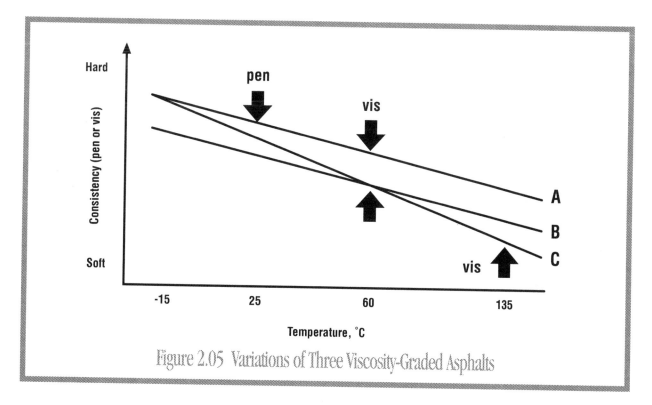

Figure 2.05 Variations of Three Viscosity-Graded Asphalts

As asphalt cement specifications evolve, limitations of previous specifications are overcome. Many asphalt tests are empirical. Field experience is required before these test results yield meaningful information. Penetration is an example of this. The penetration test represents the stiffness of the asphalt, but any relationship between asphalt penetration and performance has to be viewed through experience. Many times the correlation between test results and performance may not be very good.

Another limitation is that specifications do not illustrate the entire range of typical pavement temperatures. No low temperature properties are directly measured in the penetration and viscosity grading systems. Although viscosity is a fundamental measure of flow, it only provides information about higher temperature viscous behavior [60°C (140°F) and 135°C (275°F)]. Penetration describes only the consistency at a medium temperature [25°C (77°F)]. Low temperature behavior cannot be realistically determined from this data to predict performance.

With these limitations it is evident that penetration and viscosity asphalt specifications can classify different asphalts into the same grade. In fact these asphalts may have very different temperature and performance characteristics. Figure 2.05 shows three asphalts having the same viscosity grade because they:

- Are within the specified viscosity limits at 60°C (140°F).
- Have the minimum penetration at 25°C (77°F).
- Are above the minimum viscosity at 135°C (275°F).

While Asphalts A and B display the same temperature dependency, they have much different consistency at all temperatures. Asphalts A and C have the same consistency at -15°C but remarkably different high temperature consistency. Asphalt B has the same consistency at 60°C (140°F) but shares no other similarities with Asphalt C. Because these asphalts meet the same grade specifications, one might erroneously expect the same characteristics during construction and the same performance during hot and cold weather conditions.

Performance grading (PG) has been developed through Superpave to overcome the limitations of empirical tests and insufficient temperature characterization. The PG system uses performance-based tests to relate laboratory data to field performance. The PG tests characterize asphalts throughout the broad range of temperatures encountered in asphalt pavements.

Asphalt Specifications and Grades Asphalt cements can be graded according to four different systems:

- Penetration
- Viscosity
- Viscosity after aging
- Performance grade (PG)

The PG system has generally superceded the other systems in the U.S.

Penetration grading (ASTM D 946/AASHTO M 20) describes relative hardness based on the penetration test. In penetration grading, the higher the number the softer the asphalt. A 200-300 penetration grade asphalt is softer than a 40-50 penetration grade.

The viscosity gradings (ASTM D 3381/AASHTO M 226) use a numbering system to describe relative viscosity. The higher the number the more viscous (or thicker) the asphalt. For example, an AC-10 is softer than an AC-40.

In the viscosity after aging system (also ASTM D 3381/AASHTO M 226), asphalt is classified after it has been artificially aged. The prefix AR is used for "Aged Residue." Again, the higher the number the more viscous the material. So an AR-4000 is a softer asphalt than an AR-16000.

The performance grade, or PG system (AASHTO MP 1), describes asphalt based on the pavement temperatures under which the asphalt is expected to perform. The PG system is a part of the Superpave system. A PG 64-28 asphalt is designed for pavement temperatures as high as 64°C and as low as -28°C.

►►Testing of Asphalt Cement

This section describes, in general terms, three of the test methods for classifying asphalt. The ASTM and AASHTO references that detail the equipment and procedures related to each test are listed in Appendix D.

Penetration The *penetration* test is an empirical measure of the hardness of asphalt at room temperature. The standard penetration test (Figure 2.06) begins with conditioning a sample of asphalt cement to a temperature of 25°C (77°F) in a temperature-controlled water bath. A standard needle is then brought to bear on the surface of the asphalt under a load of 100 grams for exactly five seconds. The distance that the needle penetrates into the asphalt cement is recorded in units of 0.1 mm. The distance the needle travels is called the "penetration" of the sample.

Viscosity In the *viscosity* system, the result of the viscosity test at 60°C (140°F) is used to grade an asphalt cement. This test represents asphalt viscosity at the maximum temperature the pavement is likely to experience while in service. The result of a second viscosity test performed at 135°C (275°F) approximates the viscosity of the asphalt during mixing and laydown. Knowing the asphalt consistency at these two temperatures is the major factor in determining if the asphalt meets the specification requirements.

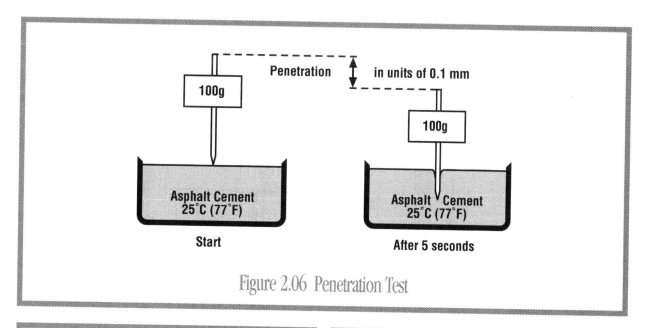

Figure 2.06 Penetration Test

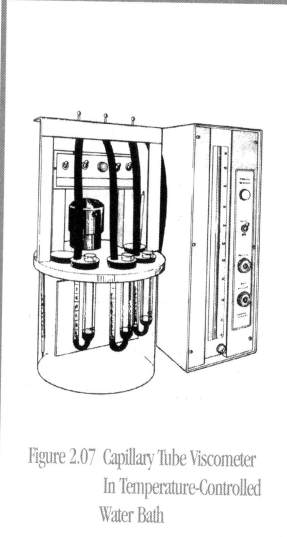

Figure 2.07 Capillary Tube Viscometer
In Temperature-Controlled
Water Bath

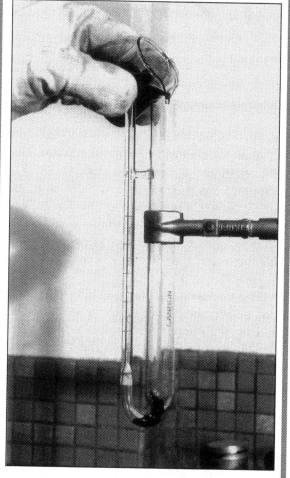

Figure 2.08 Pouring Asphalt Cement
Sample Into Viscometer

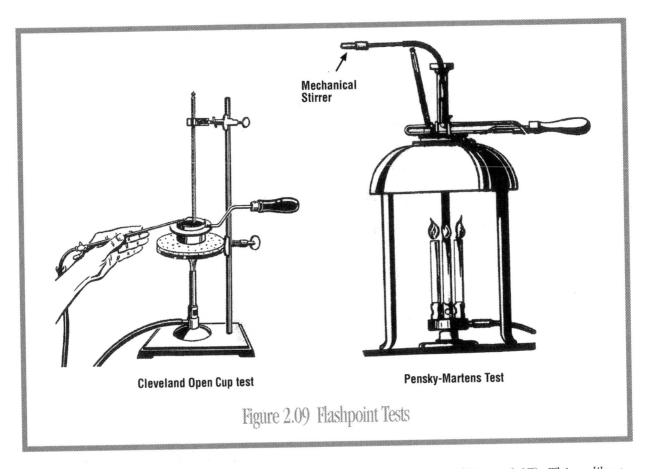

Mechanical Stirrer

Cleveland Open Cup test

Pensky-Martens Test

Figure 2.09 Flashpoint Tests

The viscosity test at 60°C (140°F) uses a capillary tube viscometer (Figure 2.07). This calibrated, glass tube measures flow of asphalt. The viscometer is mounted in a temperature controlled water bath and is preheated to 60°C (140°F). A sample of asphalt cement heated to the same temperature is then poured into the large end of the viscometer (Figure 2.08). Because asphalt cement at 60°C (140°F) is too viscous to flow readily through the narrow opening in the capillary tube, a vacuum is applied to the small end of the tube to draw the asphalt through. As the asphalt begins to flow, its progress from one mark on the tube to the next is carefully timed. This measured time is easily converted to poises, the standard unit of measurement for asphalt viscosity.

The viscosity test at 135°C (275°F) is similar to the test described above with some adaptations due to the higher temperature. First, because water boils at 100°C (212°F), a clear oil is used for the temperature-controlled bath. Second, because asphalt cement at 135°C (275°F) is fluid enough to flow through the viscometer tube, a viscometer without a vacuum is used. Third, because gravity is used to induce flow through the tube, the viscosity measurement is in centistokes instead of poises (see Appendix B for definitions).

Viscosity After Aging The *viscosity after aging* system classifies asphalt cements after they have been artificially aged using the Rolling Thin Film Oven procedure. The viscosity tests are conducted again at 60°C (140°F) and 135°C (275°F), but in accordance with Table 3 of the referenced specifications. The AR grade is determined from the viscosity test result at 60°C (140°F).

Flashpoint The *flashpoint* of asphalt cement is the lowest temperature at which volatile gases separate from a sample to "flash" in the presence of an open flame. Flashpoint must not

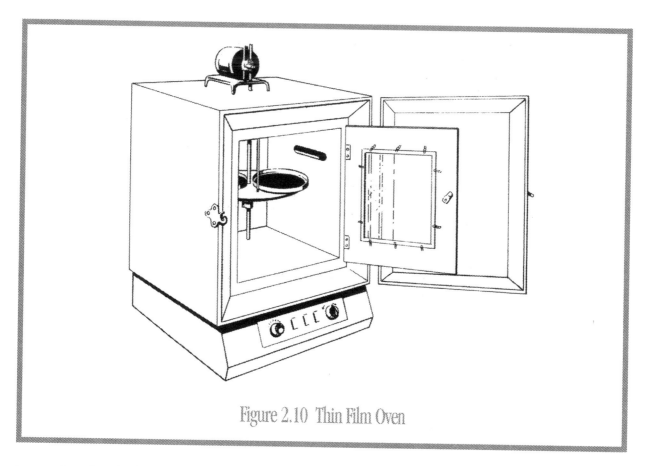

Figure 2.10 Thin Film Oven

be confused with the firepoint, the lowest temperature at which the asphalt cement will burn. Flashpoint involves only instantaneous combustion of the volatile fractions separating from the asphalt. The asphalt flashpoint is determined to identify the maximum temperature at which it can be handled and stored without danger of flashing. This is important information since asphalt cement is usually heated in storage to keep its viscosity low enough so that the material can be pumped.

The basic procedure for determining flashpoint is to gradually heat a sample of asphalt cement in a brass cup while periodically moving a small flame over the sample (Figure 2.09). The temperature at which an instantaneous flashing of vapors occurs across the surface is the flashpoint. The Cleveland Open Cup Test is the most common procedure for determining flashpoint. The Pensky-Martens Test is sometimes used.

Thin Film Oven (TFO) and Rolling Thin Film Oven (RTFO) Procedures These procedures for preparing asphalt specimens simulate conditions that occur during hot mix plant operations. Additional tests are conducted after the TFO or RTFO procedures to simulate the asphalt's properties after construction. The TFO procedure involves placing a measured sample of asphalt cement into a flat-bottomed pan so that the sample covers the pan bottom to a depth of about 3 mm (1/8-inch). The sample and pan are then placed on a rotating shelf in an oven (Figure 2.10) and kept at a temperature of 163°C (325°F) for five hours. The aged sample is then tested for its viscosity, penetration or both.

The RTFO procedure has the same purpose as the TFO test but uses different equipment and procedures. As Figure 2.11 shows, the equipment includes a 163°C (325°F) oven with a rotating vertical carriage that holds sample bottles. The asphalt sample is placed in the bottle,

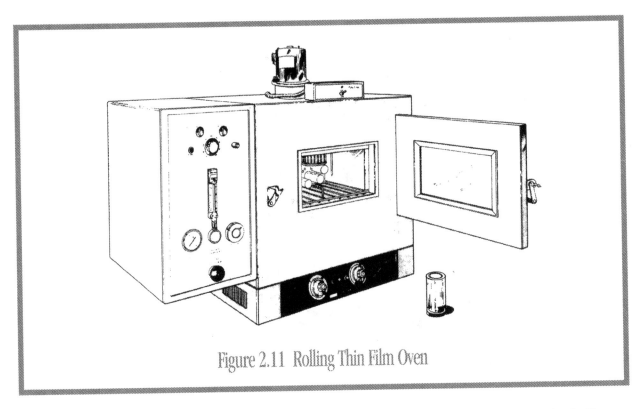

Figure 2.11 Rolling Thin Film Oven

and the bottle is placed in the carriage. The carriage rotates, continuously exposing fresh films of asphalt cement as it coats the inside of the bottle. Once during each rotation, the bottle opening passes a nozzle that blows air into the bottle.

Ductility *Ductility* is a measure of how far a sample of asphalt cement can be stretched before it breaks into two parts. It is used in the penetration and viscosity classification systems. Ductility is measured by an "extension" test in which a briquette of asphalt cement is extended, or stretched, at a specific rate and temperature (Figure 2.12). Extension is continued until the thread of asphalt cement joining the two halves of the sample breaks. The length in centimeters of the specimen at the moment it breaks is the ductility.

Solubility The *solubility* test measures the purity of asphalt cement. A sample is immersed in a solvent to dissolve the asphalt. Impurities such as salts, free carbon and nonorganic contaminants do not dissolve. These insoluble impurities are filtered out of the solution and measured as a proportion of the original sample.

Specific Gravity *Specific gravity* is the ratio of the weight of any volume of a material to the weight of an equal volume of water, both at a specified temperature. As an example, a substance with a specific gravity of 1.6 weighs 1.6 times as much as water. The specific gravity of asphalt cement is not normally indicated in the job specifications. Nonetheless, knowing the specific gravity of the asphalt cement being used is important for two reasons:

- Asphalt expands when heated and contracts when cooled. This means that the volume of a given amount of asphalt cement will be greater at higher temperatures than at lower ones. Specific gravity measurements provide a yardstick for making temperature-volume corrections, which are discussed later.

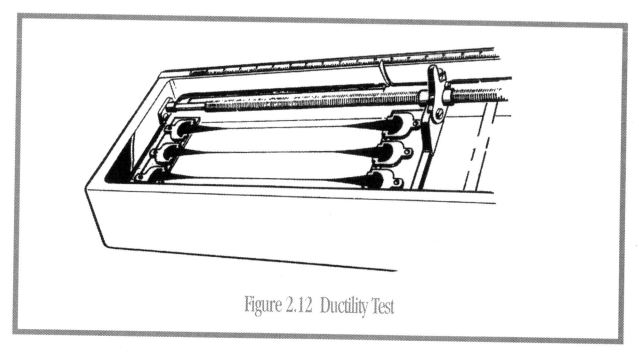

Figure 2.12 Ductility Test

- Specific gravity of the asphalt is essential in the determination of the effective asphalt content and the percentage of air voids in compacted mix specimens and compacted pavement.

Specific gravity of asphalt is usually determined using a pycnometer. Because specific gravity varies with the expansion and contraction of asphalt cement at different temperatures, results are normally expressed in terms of Sp. Gr. (Specific Gravity) at a given temperature for both the material and the water used in the test. Example: Sp. Gr. 1.05 at 15.6°/15.6°C (60°/60°F) means that the specific gravity of the asphalt cement tested is 1.05 when both the asphalt cement and the water are at 15.6°C (60°F). The specific gravity of every asphalt cement is available from the asphalt producer.

►►Superpave Asphalt Tests

The Superpave asphalt mix design system includes the performance graded (PG) asphalt binder specification. The term "asphalt binder" is used because the material can be either an unmodified or a modified asphalt cement, as long as it meets the specification criteria. The Superpave asphalt binder tests measure physical properties that can be directly related to field performance in terms of rutting, fatigue cracking and low temperature cracking. The tests are performed on asphalt at a wide range of temperatures and aging conditions with all specification criteria expressed in metric units only. Superpave characterizes asphalt at the actual pavement temperatures it will experience, and at the periods of time when distresses are most likely to occur. Details of the Superpave binder test procedures and the PG specification can be found in the Asphalt Institute's Superpave Binder manual, Superpave Series No. 1 (SP-1).

Aging Tests Superpave uses the rolling thin film oven (RTFO) and the pressure aging vessel (PAV) for aging the binder to simulate the condition of the asphalt immediately after construction (RTFO) and after years of in-service aging in the pavement (PAV). The RTFO procedure was described earlier. The PAV procedure involves placing an asphalt sample, which has already

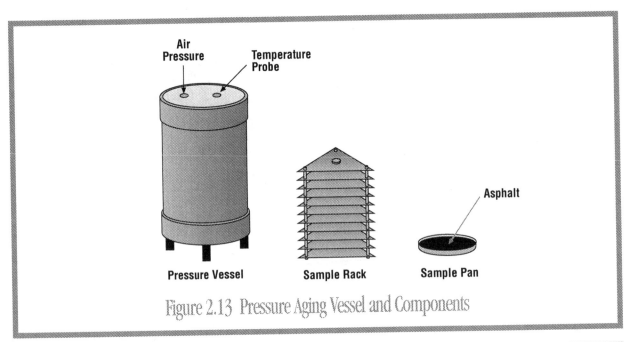

Figure 2.13 Pressure Aging Vessel and Components

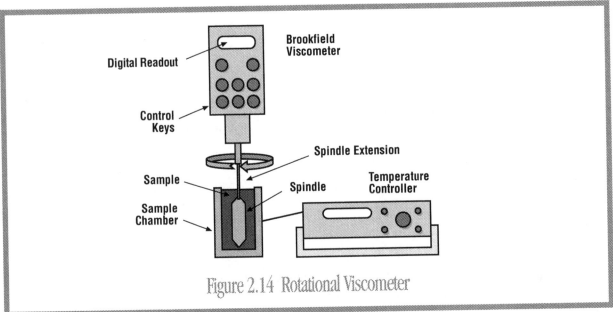

Figure 2.14 Rotational Viscometer

been aged in the RTFO procedure, inside a stainless steel pressure vessel (Figure 2.13). The vessel, heated to 100°C, is then pressurized to 2070 kPa for 20 hours. The PAV samples are then used for further testing and evaluation.

Rotational Viscometer The rotational, or Brookfield, viscometer (Figure 2.14) measures viscosity of the unaged or tank asphalt at 135°C. The test is used to determine if the asphalt is fluid enough to handle during shipment, pumping and mixing.

Dynamic Shear Rheometer As discussed earlier, asphalt is a viscoelastic material, meaning that it simultaneously behaves as an elastic material (e.g. rubber band) and a viscous material (e.g. molasses). The Dynamic Shear Rheometer (DSR) is used to characterize the viscous and elastic

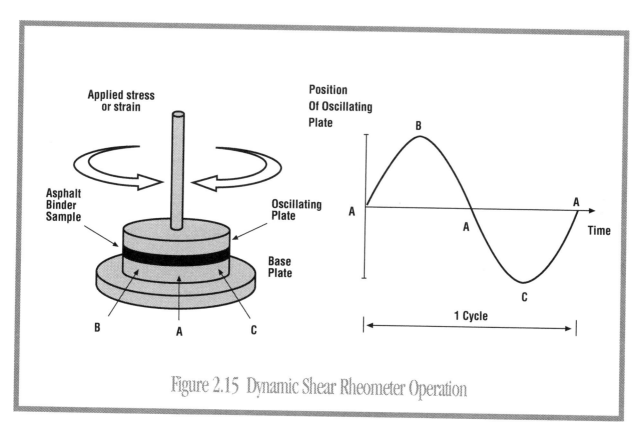

Figure 2.15 Dynamic Shear Rheometer Operation

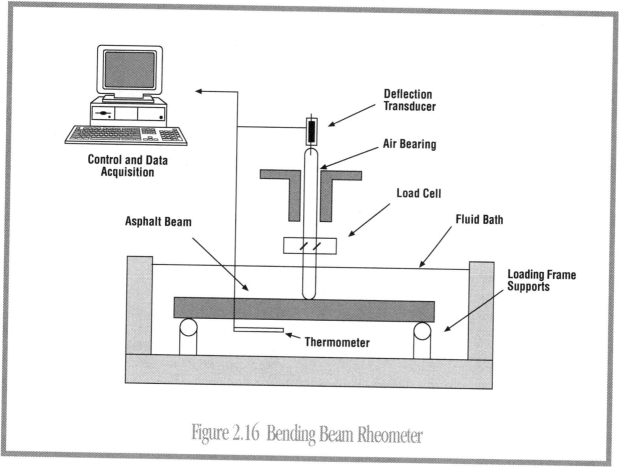

Figure 2.16 Bending Beam Rheometer

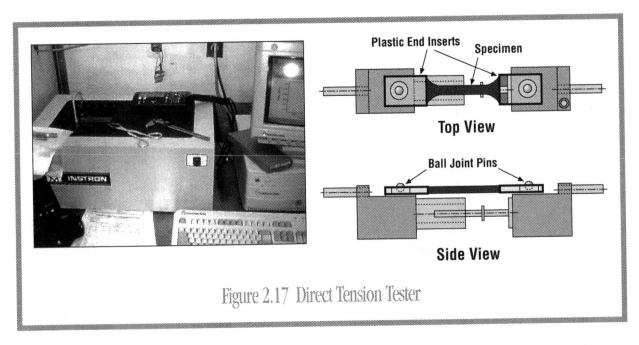

Top View

Plastic End Inserts   Specimen

Side View

Ball Joint Pins

Figure 2.17 Direct Tension Tester

behavior of asphalt binders. It does this by measuring the viscous and elastic properties of a thin binder sample sandwiched between an oscillating plate and a base plate (Figure 2.15).

The relationship between these two properties is used to measure the ability of the binder to resist permanent deformation and fatigue cracking. To resist rutting, a binder needs to be stiff and elastic; to resist fatigue cracking, the binder needs to be flexible and elastic. The balance between these two needs is a critical one.

Bending Beam Rheometer The Bending Beam Rheometer (BBR) is used to measure asphalt stiffness at very low temperatures (Figure 2.16). The test uses engineering beam theory to measure the stiffness of a small asphalt beam sample under a creep load. A creep load is used to simulate the stresses that gradually build up in a pavement when the temperature drops. The BBR test results help determine the asphalt's resistance to low temperature cracking.

Direct Tension Tester The direct tension test (Figure 2.17) measures low temperature ultimate tensile strain of an aged asphalt binder. It is performed at temperatures ranging from 0°C to -36°C, the temperature range within which asphalt exhibits brittle behavior. The direct tension test supplements the BBR in determining the asphalt's low temperature performance.

▶▶Asphalt Handling, Storage and Sampling

The safety record for handling, storing and sampling asphalt is good. Nonetheless, to prevent accidents resulting in property damage, personal injury and loss of life, everyone must know and follow good safety practices. When an accident does occur, everyone must know how to react and what first-aid treatment is appropriate. Personnel should be aware of the potential sources of contamination that might exist where asphalt is stored or handled. Proper practices must be followed to prevent contamination of samples. Finally, an understanding of the changes in volume that asphalt undergoes when heated or cooled is especially important when comparing asphalt quantities measured at different temperatures.

Safe Handling of Hot Asphalt At an asphalt plant, asphalt temperatures commonly exceed 150°C (300°F). Metal surfaces of plant equipment often range between 65°C (150°F) and 95°C (200°F). Consequently, momentary contact with hot asphalt or with plant equipment, including tanks, pipelines, dryers, boilers and boiler houses, can severely burn exposed flesh. Four general precautions against burns are:

- Be aware of where burn hazards are located.
- Use designated walk areas and stay clear of hazardous situations.
- Always wear appropriate work clothing.
- Know and follow all plant safety procedures related to handling hot material and equipment.

If a burn does occur, the general treatment guidelines are:

- In the case of localized asphalt skin burns, apply cold water or an ice pack to reduce the heat in the asphalt and the skin.
- In cases where burns cover more than 10 percent of the body (approximately the skin area of one arm or half a leg), apply lukewarm water instead of cold water. Lukewarm water will reduce the temperature of the asphalt and skin without causing shock that could be induced by applying cold water or ice to major burns.
- Do not remove the asphalt from the skin.
- Do not cover the burned area with a bandage.
- Have a physician examine the burn immediately.

Hydrogen sulfide is a product of the reaction between hydrogen and sulfur naturally present in asphalt. In low concentrations hydrogen sulfide is not dangerous; however, in the high concentrations sometimes found in storage tanks and other closed areas it can be lethal. To prevent overexposure to hydrogen sulfide fumes:

- Keep your face at least two feet away from asphalt tank hatch openings.
- Stay upwind of open hatches.
- Avoid breathing fumes when opening hatch covers or taking samples.

In case of overexposure to hydrogen sulfide fumes:
- Move victim immediately to fresh air.
- Administer oxygen if breathing is difficult.
- Start artificial respiration if breathing stops.
- Have victim examined by a physician immediately.

Storing Asphalt At a stationary asphalt plant, asphalt is stored in heated, insulated tanks with an average capacity of about 75,000 liters (20,000 gallons). Smaller, trailer-mounted tanks are used for portable plants. Their capacity is generally about one-half of stationary tanks. To keep stored asphalt fluid hot enough to be pumped readily, storage tanks are equipped with steam heating coils, hot oil coils or gas-fired or electric heaters. Certain precautions regarding tank temperatures must be followed to ensure safety:

- Tank temperatures should be checked and recorded regularly. When measuring tank temperatures, an instrument designed for that purpose should be used.
- Taking temperature readings near the heating coils, tank shell or tank bottom should be avoided. Such readings are generally inaccurate indications of the overall asphalt temperature.

Table 2.01 Guideline Temperatures for Storing Asphalts

Type & Grade	Reference Specification	Minimum Flash °C (°F)	Storage Temperature °C (°F)
PG 46, 52, 58, 64, 70, 76, 82	AASHTO MP 1	230	*
AC-2.5	AASHTO M 226	163 (325)	160 (320)
-5	ASTM D 3381	177 (350)	166 (330)
-10		219 (425)	174 (345)
-20, -30, -40		232 (450)	177 (350)
AR-1000	AASHTO M 226	205 (400)	163 (325)
-2000	ASTM D 3381	219 (425)	168 (335)
-4000		227 (440)	177 (350)
-8000		232 (450)	177 (350)
-16000		238 (460)	177 (350)
Pen 40-50, 60-70,	AASHTO M 20	232 (450)	177 (350)
85-100	ASTM D 946		
120-150		219 (425)	177 (350)
200-300		177 (350)	168 (335)
RC-70	AASHTO M 81	——-	71 (160)
-250	ASTM D 2028	27 (80)	91 (195)
-800		27 (80)	99 (210)
-3000		27 (80)	99 (210)
MC-30	AASHTO M 82	38 (100)	54 (130)
-70	ASTM D 2027	38 (100)	71 (160)
-250		66 (150)	91 (195)
-800		66 (150)	99 (210)
-3000		66 (150)	99 (210)
RS-1	AASHTO M 140 & M 208	——-	50 (122)
RS-2, HFRS-2, CRS-1, CRS-2	ASTM D 977 & D 2397	——-	75 (167)
SS-1, SS-1h, CSS-1, CSS-1h, MS-1, HFMS-1		——-	45 (113)
MS-2, MS-2h, HFMS-2, HFMS-2h, HFMS-2s, CMS-2, CMS-2h		——-	75 (167)

* For Performance Grade Asphalt Binders, use the storage temperatures recommended by the asphalt producer or supplier.

- Tank storage temperatures for asphalt binders are typically kept close to the desired field mixing temperature, generally in the range of 150°C to 175°C (302°F to 347°F). This mixing temperature, at which the asphalt binder and aggregate are mixed, is generally established after determining the desired field compaction temperature. Because the mix will cool slightly while being stored and hauled to the job site, the mixing temperature will be set slightly higher than the compaction temperature. The asphalt binder producer or supplier should be consulted for guidelines in establishing these mixing and compaction temperatures. Binder storage and mixing temperatures should not exceed 175°C (347°F). Table 2.01 presents temperature guidelines for storing various types and grades of asphalt.
- Check storage tanks, and coils should be checked regularly for signs of damage or leakage.

Table 2.02 Guide for Loading Asphalt Products

LAST PRODUCT IN TANK	PRODUCT TO BE LOADED			
	Asphalt Cement/Binder	Cutback Asphalt	Cationic Emulsion	Anionic Emulsion
Asphalt Cement/Binder	OK to load	OK to load	Empty to no measurable quantity	Empty to no measurable quantity
Cutback Asphalt	Empty *	OK to load	Empty to no measurable quantity	Empty to no measurable quantity
Cationic Emulsion	Empty *	Empty to no measurable quantity	OK to load	Empty to no measurable quantity
Anionic Emulsion	Empty *	Empty to no measurable quantity	Empty to no measurable quantity	OK to load
Crude Petroleum and Residual Fuel Oils	Empty *	Empty to no measurable quantity	Empty to no measurable quantity	Empty to no measurable quantity
Any Product Not Listed Above	Tank must be cleaned	Tank must be cleaned	Tank must be cleaned	Tank must be cleaned

* Any material remaining will produce dangerous conditions

Trucks and railroad tank cars that normally carry asphalt may be used to transport other petroleum products. Contamination can cause an asphalt to not meet specifications, as well as an increased danger of fire or explosion. To minimize the potential hazards, the guidelines in Table 2.02 should be followed when loading asphalt products.

Sampling Asphalt The only way to know whether asphalt delivered to the plants meets specifications is to take samples of the material and test the samples in a laboratory. For meaningful test results, the samples must be representative of the entire shipment. Contamination or other alteration of the sample before testing is likely to produce misleading test results. Such results could be used to reject an entire shipment of asphalt cement even though it does meet specifications. Details on sampling asphalt are contained in Chapter 4.

▶▶ Asphalt Temperature-Volume Relationships

Asphalt expands when heated and contracts when cooled. These changes in volume can cause confusion because regardless of the temperature at which asphalt is shipped and stored, its volume at 15°C (60°F) is the basis for payments and project records. Consequently, when a shipment of asphalt is delivered at 150°C (300°F), its volume at 15°C (60°F) must be calculated and recorded.

The calculation requires two items of information:

- The temperature of the asphalt.
- The specific gravity of the asphalt.

The asphalt's temperature and specific gravity are used to locate the proper correction factor in Table 2.03. These tables have been in use for at least four decades and are the only data currently available for temperature corrections above 150°C (300°F). Nonetheless, the accuracy of the tables is not guaranteed. The asphalt's temperature and the necessary correction factor are used in this formula to calculate its volume at 15°C (60°F):

$$V = V_t(CF)$$

where,

V = Volume at 15°C (60°F)
V_t = Volume at given temperature
CF = Correction Factor from Table 2.03

EXAMPLE: A truck has just delivered 19,000 liters (5020 gallons) of asphalt at a temperature of 150°C (300°F). The specific gravity of the asphalt is 0.970. What would the asphalt's volume be at 15°C (60°F)? Because its specific gravity is at or above 0.967, the Column A factors (Table 2.03) are used to find the correction factor. For 150°C, the correction factor listed is 0.9177.

Therefore,

V = 19,000 liters x 0.9177 *OR 5020 gallons* x 0.9177
  = 17,436 liters *OR 4607 gallons*

Aggregate

Aggregate, also referred to as rock, granular material, and mineral aggregate, is any hard, inert mineral material used in hot mix asphalt. Typical aggregates include sand, gravel, crushed stone, slag and rock dust. Aggregate makes up 90-95 percent by weight and 75-85 percent by volume of most HMA. Because it provides most of the load-bearing characteristics, pavement performance is heavily influenced by the choice of a proper aggregate for a particular job.

►►Aggregate Classification

Rock is divided into three general types: sedimentary, igneous and metamorphic (Table 2.04). These classifications are based upon the way in which each type is formed.

Sedimentary Sedimentary rocks are formed in layers by the accumulation of sediment (fine particles) that is deposited by wind and water. Sediment may contain:

- Mineral particles or fragments (as in the case of sandstone and shale).
- Remains or products of animals (certain limestones).
- Plants (coal).

Table 2.03 Temperature-Volume Corrections for Asphalt

Observed Temperature, °C	Volume Correction Factor to 15°C [2] A	B	Observed Temperature, °C	Volume Correction Factor to 15°C [2] A	B	Observed Temperature, °C	Volume Correction Factor to 15°C [2] A	B	Observed Temperature, °C	Volume Correction Factor to 15°C [2] A	B
-25.0	1.0254	1.0290	12.5	1.0016	1.0018	50.0	0.9782	0.9752	87.5	0.9552	0.9492
-24.5	1.0251	1.0286	13.0	1.0012	1.0014	50.5	0.9779	0.9749	88.0	0.9548	0.9489
-24.0	1.0248	1.0283	13.5	1.0009	1.0014	51.0	0.9776	0.9745	88.5	0.9545	0.9485
-23.5	1.0244	1.0279	14.0	1.0006	1.0007	51.5	0.9973	0.9742	89.0	0.9542	0.9482
-23.0	1.0241	1.0276	14.5	1.0003	1.0004	52.0	0.9770	0.9738	89.5	0.9539	0.9478
-22.5	1.0238	1.0272	15.0	1.0000	1.0000	52.5	0.9767	0.9735	90.0	0.9536	0.9475
-22.0	1.0235	1.0268	15.5	0.9997	0.9998	53.0	0.9763	0.9731	90.5	0.9533	0.9472
-21.5	1.0232	1.0265	16.0	0.9994	0.9993	53.5	0.9760	0.9728	91.0	0.9530	0.9468
-21.0	1.0228	1.0261	16.5	0.9991	0.9989	54.0	0.9757	0.9724	91.5	0.9527	0.9465
-20.5	1.0225	1.0258	17.0	0.9988	0.9986	54.5	0.9754	0.9721	92.0	0.9524	0.9461
-20.0	1.0222	1.0254	17.5	0.9985	0.9982	55.0	0.9751	0.9717	92.5	0.9521	0.9458
-19.5	1.0219	1.0250	18.0	0.9981	0.9978	55.5	0.9748	0.9714	93.0	0.9518	0.9455
-19.0	1.0216	1.0247	18.5	0.9978	0.9975	56.0	0.9745	0.9710	93.5	0.9515	0.9451
-18.5	1.0212	1.0243	19.0	0.9975	0.9971	56.5	0.9742	0.9707	94.0	0.9512	0.9448
-18.0	1.0209	1.0239	19.5	0.9972	0.9968	57.0	0.9739	0.9703	94.5	0.9509	0.9444
-17.5	1.0206	1.0236	20.0	0.9969	0.9964	57.5	0.9736	0.9700	95.0	0.9506	0.9441
-17.0	1.0203	1.0232	20.5	0.9966	0.9961	58.0	0.9732	0.9696	95.5	0.9503	0.9438
-16.5	1.0200	1.0228	21.0	0.9963	0.9957	58.5	0.9729	0.9693	96.0	0.9500	0.9434
-16.0	1.0196	1.0224	21.5	0.9959	0.9954	59.0	0.9726	0.9689	96.5	0.9497	0.9431
-15.5	1.0193	1.0221	22.0	0.9956	0.9950	59.5	0.9723	0.9686	97.0	0.9494	0.9427
-15.0	1.0190	1.0217	22.5	0.9953	0.9947	60.0	0.9720	0.9682	97.5	0.9491	0.9424
-14.5	1.0187	1.0213	23.0	0.9950	0.9943	60.5	0.9717	0.9679	98.0	0.9488	0.9421
-14.0	1.0184	1.0210	23.5	0.9947	0.9940	61.0	0.9714	0.9675	98.5	0.9485	0.9417
-13.5	1.0180	1.0206	24.0	0.9943	0.9936	61.5	0.9711	0.9672	99.0	0.9482	0.9414
-13.0	1.0177	1.0203	24.5	0.9940	0.9933	62.0	0.9708	0.9668	99.5	0.9479	0.9410
-12.5	1.0174	1.0199	25.0	0.9937	0.9929	62.5	0.9705	0.9665	100.0	0.9476	0.9407
-12.0	1.0171	1.0195	25.5	0.9934	0.9925	63.0	0.9701	0.9661	100.5	0.9473	0.9404
-11.5	1.0168	1.0192	26.0	0.9931	0.9922	63.5	0.9698	0.9658	101.0	0.9470	0.9400
-11.0	1.0164	1.0188	26.5	0.9928	0.9918	64.0	0.9695	0.9654	101.5	0.9467	0.9397
-10.5	1.0161	1.0185	27.0	0.9925	0.9915	64.5	0.9692	0.9651	102.0	0.9464	0.9393
-10.0	1.0158	1.0181	27.5	0.9922	0.9911	65.0	0.9689	0.9647	102.5	0.9461	0.9390
-9.5	1.0155	1.0177	28.0	0.9918	0.9907	65.5	0.9686	0.9644	103.0	0.9458	0.9387
-9.0	1.0152	1.0174	28.5	0.9915	0.9904	66.0	0.9683	0.9640	103.5	0.9455	0.9383
-8.5	1.0148	1.0170	29.0	0.9912	0.9900	66.5	0.9680	0.9637	104.0	0.9452	0.9380
-8.0	1.0145	1.0166	29.5	0.9909	0.9897	67.0	0.9677	0.9633	104.5	0.9449	0.9376
-7.5	1.0142	1.0163	30.0	0.9906	0.9893	67.5	0.9674	0.9630	105.0	0.9446	0.9373
-7.0	1.0139	1.0159	30.5	0.9903	0.9889	68.0	0.9670	0.9626	105.5	0.9443	0.9370
-6.5	1.0136	1.0155	31.0	0.9900	0.9886	68.5	0.9667	0.9623	106.0	0.9440	0.9366
-6.0	1.0132	1.0151	31.5	0.9897	0.9882	69.0	0.9664	0.9619	106.5	0.9437	0.9363
-.5.5	1.0129	1.0148	32.0	0.9894	0.9879	69.5	0.9661	0.9616	107.0	0.9434	0.9359
-5.0	1.0126	1.0144	32.5	0.9891	0.9875	70.0	0.9658	0.9612	107.5	0.9431	0.9356
-4.5	1.0123	1.0140	33.0	0.9887	0.9871	70.5	0.9655	0.9609	108.0	0.9428	0.9353
-4.0	1.0120	1.0137	33.5	0.9884	0.9868	71.0	0.9652	0.9605	108.5	0.9425	0.9349
-3.5	1.0117	1.0133	34.0	0.9881	0.9864	71.5	0.9649	0.9602	109.0	0.9422	0.9346
-3.0	1.0114	1.0130	34.5	0.9878	0.9861	72.0	0.9646	0.9598	109.5	0.9419	0.9342
-2.5	1.0111	1.0126	35.0	0.9875	0.9857	72.5	0.9643	0.9595	110.0	0.9416	0.9339
-2.0	1.0107	1.0122	35.5	0.9872	0.9854	73.0	0.9640	0.9592	110.5	0.9413	0.9336
-1.5	1.0104	1.0119	36.0	0.9869	0.9850	73.5	0.9637	0.9588	111.0	0.9410	0.9332
-1.0	1.0101	1.0115	36.5	0.9866	0.9847	74.0	0.9634	0.9585	111.5	0.9407	0.9329
-0.5	1.0098	1.0112	37.0	0.9863	0.9843	74.5	0.9631	0.9581	112.0	0.9404	0.9325
0	1.0095	1,0108	37.5	0.9860	0.9840	75.0	0.9628	0.9578	112.5	0.9401	0.9322
0.5	1.0092	1.0104	38.0	0.9856	0.9836	75.5	0.9625	0.9575	113.0	0.9397	0.9319
1.0	1.0089	1.0101	38.5	0.9853	0.9833	76.0	0.9622	0.9571	113.5	0.9394	0.9315
1.5	1.0085	1.0097	39.0	0.9850	0.9829	76.5	0.9619	0.9568	114.0	0.9391	0.9312
2.0	1.0082	1.0094	39.5	0.9847	0.9826	77.0	0.9616	0.9564	114.5	0.9388	0.9308
2.5	1.0079	1.0090	40.0	0.9844	0.9822	77.5	0.9613	0.9561	115.0	0.9385	0.9305
3.0	1.0076	1.0086	40.5	0.9841	0.9819	78.0	0.9609	0.9557	115.5	0.9382	0.9302
3.5	1.0073	1.0083	41.0	0.9838	0.9815	78.5	0.9606	0.9554	116.0	0.9379	0.9298
4.0	1.0069	1.0079	41.5	0.9835	0.9812	79.0	0.9603	0.9550	116.5	0.9376	0.9295
4.5	1.0066	1.0076	42.0	0.9832	0.9808	79.5	0.9600	0.9547	117.0	0.9373	0.9292
5.0	1.0063	1.0072	42.5	0.9829	0.9805	80.0	0.9597	0.9543	117.5	0.9371	0.9289
5.5	1.0060	1.0068	43.0	0.9825	0.9801	80.5	0.9594	0.9540	118.0	0.9368	0.9285
6.0	1.0057	1.0065	43.5	0.9822	0.9798	81.0	0.9591	0.9536	118.5	0.9365	0.9282
6.5	1.0053	1.0061	44.0	0.9819	0.9794	81.5	0.9588	0.9533	119.0	0.9362	0.9279
7.0	1.0050	1.0058	44.5	0.9816	0.9791	82.0	0.9585	0.9529	119.5	0.9359	0.9275
7.5	1.0047	1.0054	45.0	0.9813	0.9787	82.5	0.9582	0.9526	120.0	0.9356	0.9272
8.0	1.0044	1.0050	45.5	0.9810	0.9784	83.0	0.9578	0.9523	120.5	0.9353	0.9269
8.5	1.0041	1.0047	46.0	0.9807	0.9780	83.5	0.9576	0.9519	121.0	0.9350	0.9265
9.0	1.0037	1.0043	46.5	0.9804	0.9777	84.0	0.9573	0.9516	121.5	0.9347	0.9262
9.5	1.0034	1.0040	47.0	0.9801	0.9973	84.5	0.9570	0.9512	122.0	0.9344	0.9258
10.0	1.0031	1.0036	47.5	0.9798	0.9770	85.0	0.9567	0.9509	122.5	0.9341	0.9255
10.5	1.0028	1.0032	48.0	0.9794	0.9766	85.5	0.9564	0.9506	123.0	0.9338	0.9252
11.0	1.0025	1.0029	48.5	0.9791	0.9763	86.0	0.9561	0.9502	123.5	0.9335	0.9248
11.5	1.0022	1.0025	49.0	0.9788	0.9759	86.5	0.9558	0.9499	124.0	0.9332	0.9245
12.0	1.0019	1.0022	49.5	0.9785	0.9756	87.0	0.9555	0.9495	124.5	0.9329	0.9241

[2] Use column A factors for asphalts with specific gravity at 15° C of 0.967 or higher. Use column B factors for asphalts with specific gravity at 15° C of 0.850 to 0.966.

Table 2.03 Temperature-Volume Corrections for Asphalt (continued)

Observed Temperature, °C	Volume Correction Factor to 15°C [2] A	B	Observed Temperature, °C	Volume Correction Factor to 15°C [2] A	B	Observed Temperature, °C	Volume Correction Factor to 15°C [2] A	B	Observed Temperature, °C	Volume Correction Factor to 15°C [2] A	B
125.0	0.9326	0.9238	162.5	0.9104	0.8991	200.0	0.8886	0.8749	237.5	0.8673	0.8514
125.5	0.9323	0.9235	163.0	0.9101	0.8987	200.5	0.8883	0.8746	238.0	0.8670	0.8510
126.0	0.9320	0.9231	163.5	0.9098	0.8984	201.0	0.8880	0.8743	238.5	0.8667	0.8507
126.5	0.9317	0.9228	164.0	0.9095	0.8981	201.5	0.8877	0.8739	239.0	0.8664	0.8504
127.0	0.9314	0.9225	164.5	0.9092	0.8977	202.0	0.8874	0.8736	239.5	0.8661	0.8501
127.5	0.9311	0.9222	165.0	0.9089	0.8974	202.5	0.8872	0.8733	240.0	0.8658	0.8498
128.0	0.9308	0.9218	165.5	0.9086	0.8971	203.0	0.8869	0.8730	240.5	0.8655	0.8495
128.5	0.9305	0.9215	166.0	0.9083	0.8968	203.5	0.8866	0.8727	241.0	0.8652	0.8492
129.0	0.9302	0.9212	166.5	0.9080	0.8964	204.0	0.8863	0.8723	241.5	0.8650	0.8489
129.5	0.9299	0.9208	167.0	0.9077	0.8961	204.5	0.8860	0.8720	242.0	0.8647	0.8486
130.0	0.9296	0.9205	167.5	0.9075	0.8958	205.0	0.8857	0.8717	242.5	0.8644	0.8483
130.5	0.9293	0.9202	168.0	0.9072	0.8955	205.5	0.8854	0.8714	243.0	0.8641	0.8480
131.0	0.9290	0.9198	168.5	0.9069	0.8952	206.0	0.8851	0.8711	243.5	0.8638	0.8477
131.5	0.9287	0.9195	169.0	0.9066	0.8948	206.5	0.8849	0.8708	244.0	0.8636	0.8474
132.0	0.9284	0.9191	169.5	0.9063	0.8945	207.0	0.8846	0.8705	244.5	0.8633	0.8471
132.5	0.9281	0.9188	170.0	0.9060	0.8942	207.5	0.8843	0.8702	245.0	0.8630	0.8468
133.0	0.9278	0.9185	170.5	0.9057	0.8939	208.0	0.8840	0.8698	245.5	0.8627	0.8465
133.5	0.9275	0.9181	171.0	0.9054	0.8935	208.5	0.8837	0.8695	246.0	0.8624	0.8462
134.0	0.9272	0.9178	171.5	0.9051	0.8932	209.0	0.8835	0.8692	246.5	0.8622	0.8459
134.5	0.9269	0.9174	172.0	0.9048	0.8929	209.5	0.8832	0.8689	247.0	0.8619	0.8456
135.0	0.9266	0.9171	172.5	0.9046	0.8926	210.0	0.8829	0.8686	247.5	0.8616	0.8453
135.5	0.9263	0.9168	173.0	0.9043	0.8922	210.5	0.8826	0.8683	248.0	0.8613	0.8449
136.0	0.9260	0.9164	173.5	0.9040	0.8919	211.0	0.8823	0.8680	248.5	0.8610	0.8446
136.5	0.9257	0.9161	174.0	0.9037	0.8916	211.5	0.8820	0.8676	249.0	0.8608	0.8443
137.0	0.9254	0.9158	174.5	0.9034	0.8912	212.0	0.8817	0.8673	249.5	0.8605	0.8440
137.5	0.9251	0.9155	175.0	0.9031	0.8909	212.5	0.8815	0.8670	250.0	0.8602	0.8437
138.0	0.9248	0.9151	175.5	0.9028	0.8906	213.0	0.8812	0.8667	250.5	0.8599	0.8434
138.5	0.9245	0.9148	176.0	0.9025	0.8903	213.5	0.8809	0.8664	251.0	0.8596	0.8431
139.0	0.9242	0.9145	176.5	0.9022	0.8899	214.0	0.8806	0.8660	251.5	0.8594	0.8428
139.5	0.9239	0.9141	177.0	0.9019	0.8896	214.5	0.8803	0.8657	252.0	0.8591	0.8425
140.0	0.9236	0.9138	177.5	0.9017	0.8993	215.0	0.8800	0.8654	252.5	0.8588	0.8422
140.5	0.9233	0.9135	178.0	0.9014	0.8990	215.5	0.8797	0.8651	253.0	0.8585	0.8418
141.0	0.9230	0.9131	178.5	0.9011	0.8887	216.0	0.8794	0.8648	253.5	0.8582	0.8415
141.5	0.9227	0.9128	179.0	0.9008	0.8883	216.5	0.8792	0.8645	254.0	0.8580	0.8412
142.0	0.9224	0.9125	179.5	0.9005	0.8880	217.0	0.8789	0.8642	254.5	0.8577	0.8409
142.5	0.9222	0.9122	180.0	0.9002	0.8877	217.5	0.8786	0.8639	255.0	0.8574	0.8406
143.0	0.9219	0.9118	180.5	0.8999	0.8874	218.0	0.8783	0.8635	255.5	0.8571	0.8403
143.5	0.9216	0.9115	181.0	0.8996	0.8871	218.5	0.8780	0.8632	256.0	0.8586	0.8400
144.0	0.9213	0.9112	181.5	0.8993	0.8867	219.0	0.8778	0.8629	256.5	0.8566	0.8397
144.5	0.9210	0.9108	182.0	0.8990	0.8864	219.5	0.8775	0.8626	257.0	0.8563	0.8394
145.0	0.9207	0.9105	182.5	0.8988	0.8861	220.0	0.8772	0.8623	257.5	0.8560	0.8391
145.5	0.9204	0.9102	183.0	0.8985	0.8858	220.5	0.8769	0.8620	258.0	0.8557	0.8388
146.0	0.9201	0.9098	183.5	0.8982	0.8855	221.0	0.8766	0.8617	258.5	0.8554	0.8385
146.5	0.9198	0.9095	184.0	0.8979	0.8851	221.5	0.8763	0.8614	259.0	0.8552	0.8382
147.0	0.9195	0.9092	184.5	0.8976	0.8848	222.0	0.8760	0.8611	259.5	0.8549	0.8379
147.5	0.9192	0.9089	185.0	0.8973	0.8845	222.5	0.8758	0.8606	260.0	0.8546	0.8376
148.0	0.9189	0.9085	185.5	0.8970	0.8842	223.0	0.8755	0.8604	260.5	0.8543	0.8373
148.5	0.9186	0.9082	186.0	0.8967	0.8839	223.5	0.8752	0.8601	261.0	0.8540	0.8370
149.0	0.9183	0.9079	186.5	0.8964	0.8835	224.0	0.8749	0.8598	261.5	0.8538	0.8367
149.5	0.9180	0.9075	187.0	0.8961	0.8832	224.5	0.8746	0.8595	262.0	0.8535	0.8364
150.0	0.9177	0.9072	187.5	0.8959	0.8829	225.0	0.8743	0.8592	262.5	0.8532	0.8361
150.5	0.9174	0.9069	188.0	0.8956	0.8826	225.5	0.8740	0.8589	263.0	0.8529	0.8357
151.0	0.9171	0.9065	188.5	0.8953	0.8823	226.0	0.8737	0.8586	263.5	0.8526	0.8354
151.5	0.9168	0.9062	189.0	0.8950	0.8819	226.5	0.8735	0.8582	264.0	0.8524	0.8351
152.0	0.9165	0.9059	189.5	0.8947	0.8816	227.0	0.8732	0.8579	264.5	0.8521	0.8348
152.5	0.9163	0.9056	190.0	0.8944	0.8813	227.5	0.8729	0.8576	265.0	0.8518	0.8345
153.0	0.9160	0.9052	190.5	0.8941	0.8810	228.0	0.8726	0.8573	265.5	0.8515	0.8342
153.3	0.9157	0.9049	191.0	0.8938	0.8807	228.5	0.8723	0.8570	266.0	0.8512	0.8339
154.0	0.9154	0.9046	191.5	0.8935	0.8803	229.0	0.8721	0.8566	266.5	0.8510	0.8336
154.5	0.9151	0.9042	192.0	0.8932	0.8800	229.5	0.8718	0.8563	267.0	0.8507	0.8333
155.0	0.9148	0.9039	192.5	0.8930	0.8797	230.0	0.8715	0.8560	267.5	0.8504	0.8330
155.5	0.9145	0.9036	193.0	0.8927	0.8794	230.5	0.8712	0.8557	268.0	0.8501	0.8326
156.0	0.9142	0.9033	193.5	0.8924	0.8791	231.0	0.8709	0.8554	268.5	0.8498	0.8323
156.5	0.9139	0.9029	194.0	0.8921	0.8787	231.5	0.8707	0.8551	269.0	0.8496	0.8320
157.0	0.9136	0.9026	194.5	0.8918	0.8784	232.0	0.8704	0.8548	269.5	0.8493	0.8317
157.5	0.9133	0.9023	195.0	0.8915	0.8781	232.5	0.8701	0.8545	270.0	0.8490	0.8314
158.0	0.9130	0.9020	195.5	0.8912	0.8778	233.0	0.8698	0.8541	270.5	0.8487	0.8311
158.5	0.9127	0.9017	196.0	0.8909	0.8775	233.5	0.8695	0.8538	271.0	0.8484	0.8308
159.0	0.9124	0.9013	196.5	0.8906	0.8771	234.0	0.8693	0.8535	271.5	0.8482	0.8305
159.5	0.9121	0.9010	197.0	0.8903	0.8768	234.5	0.8690	0.8532	272.0	0.8479	0.8302
160.0	0.9118	0.9007	197.5	0.8901	0.8765	235.0	0.8687	0.8529	272.5	0.8476	0.8299
160.5	0.9115	0.9004	198.0	0.8898	0.8762	235.5	0.8684	0.8526	273.0	0.8473	0.8296
161.0	0.9112	0.9000	198.5	0.8895	0.8759	236.0	0.8681	0.8523	273.5	0.8470	0.8293
161.5	0.9109	0.8997	199.0	0.8892	0.8755	236.5	0.8678	0.8520	274.0	0.8468	0.8290
162.0	0.9106	0.8994	199.5	0.8889	0.8752	237.0	0.8675	0.8517	274.5	0.8465	0.8287

[2] Use column A factors for asphalts with specific gravity at 15° C of 0.967 or higher. Use column B factors for asphalts with specific gravity at 15° C of 0.850 to 0.966.

Table 2.03 Temperature-Volume Corrections for Asphalt (continued)

Observed Temperature, °F	Volume Correction Factor to 60° F [1] A	B	Observed Temperature, °F	Volume Correction Factor to 60° F [1] A	B	Observed Temperature, °F	Volume Correction Factor to 60° F [1] A	B	Observed Temperature, °F	Volume Correction Factor to 60° F [1] A	B
0	1.0211	1.0241	70	0.9965	0.9950	140	0.9723	0.9686	210	0.9486	0.9418
1	1.0208	1.0237	71	0.9962	0.9956	141	0.9720	0.9682	211	0.9483	0.9414
2	1.0204	1.0233	72	0.9958	0.9952	142	0.9716	0.9678	212	0.9479	0.9410
3	1.0201	1.0229	73	0.9955	0.9948	143	0.9713	0.9674	213	0.9476	0.9407
4	1.0197	1.0225	74	0.9951	0.9944	144	0.9710	0.9670	214	0.9472	0.9403
5	1.0194	1.0221	75	0.9948	0.9940	145	0.9706	0.9666	215	0.9469	0.9399
6	1.0190	1.0217	76	0.9944	0.9936	146	0.9703	0.9662	216	0.9466	0.9395
7	1.0186	1.0213	77	0.9941	0.9932	147	0.9699	0.9659	217	0.9462	0.9391
8	1.0183	1.0209	78	0.9937	0.9929	148	0.9696	0.9655	218	0.9459	0.9388
9	1.0179	1.0205	79	0.9934	0.9925	149	0.9693	0.9651	219	0.9456	0.9386
10	1.0176	1.0201	80	0.9930	0.9921	150	0.9689	0.9647	220	0.9452	0.9380
11	1.0172	1.0197	81	0.9927	0.9917	151	0.9686	0.9643	221	0.9449	0.9376
12	1.0169	1.0193	82	0.9923	0.9913	152	0.9682	0.9639	222	0.9446	0.9373
13	1.0165	1.0189	83	0.9920	0.9909	153	0.9679	0.9635	223	0.9442	0.9369
14	1.0162	1.0185	84	0.9916	0.9905	154	0.9675	0.9632	224	0.9439	0.9365
15	1.0158	1.0181	85	0.9913	0.9901	155	0.9672	0.9628	225	0.9436	0.9361
16	1.0155	1.0177	86	0.9909	0.9897	156	0.9669	0.9624	226	0.9432	0.9358
17	1.0151	1.0173	87	0.9909	0.9893	157	0.9665	0.9620	227	0.9429	0.9354
18	1.0148	1.0168	88	0.9902	0.9889	158	0.9662	0.9616	228	0.9426	0.9350
19	1.0144	1.0164	89	0.9899	0.9885	159	0.9658	0.9612	229	0.9422	0.9346
20	1.0141	1.0160	90	0.9896	0.9881	160	0.9655	0.9609	230	0.9419	0.9343
21	1.0137	1.0156	91	0.9892	0.9877	161	0.9652	0.9605	231	0.9416	0.9339
22	1.0133	1.0152	92	0.9889	0.9873	162	0.9648	0.9601	232	0.9412	0.9335
23	1.0130	1.0148	93	0.9885	0.9869	163	0.9645	0.9597	233	0.9409	0.9331
24	1.0126	1.0144	94	0.9882	0.9865	164	0.9641	0.9593	234	0.9405	0.9328
25	1.0123	1.0140	95	0.9878	0.9861	165	0.9638	0.9589	235	0.9402	0.9324
26	1.0119	1.0136	96	0.9875	0.9857	166	0.9835	0.9585	236	0.9399	0.9320
27	1.0116	1.0132	97	0.9871	0.9854	167	0.9631	0.9582	237	0.9395	0.9316
28	1.0112	1.0128	98	0.9868	0.9850	168	0.9628	0.9578	238	0.9392	0.9313
29	1.0109	1.0124	99	0.9864	0.9846	169	0.9624	0.9574	239	0.9389	0.9309
30	1.0105	1.0120	100	0.9861	0.9842	170	0.9621	0.9570	240	0.9385	0.9305
31	1.0102	1.0116	101	0.9857	0.9838	171	0.9618	0.9566	241	0.9382	0.9301
32	1.0098	1.0112	102	0.9854	0.9834	172	0.9614	0.9562	242	0.9379	0.9298
33	1.0095	1.0108	103	0.9851	0.9830	173	0.9611	0.9559	243	0.9375	0.9294
34	1.0091	1.0104	104	0.9847	0.9826	174	0.9607	0.9555	244	0.9372	0.9290
35	1.0088	1.0100	105	0.9844	0.9822	175	0.9604	0.9551	245	0.9369	0.9286
36	1.0084	1.0096	106	0.9840	0.9818	176	0.9601	0.9547	246	0.9365	0.9283
37	1.0081	1.0092	107	0.9887	0.9814	177	0.9597	0.9543	247	0.9362	0.9279
38	1.0077	1.0088	108	0.9833	0.9810	178	0.9594	0.9539	248	0.9359	0.9275
39	1.0074	1.0084	109	0.9830	0.9806	179	0.9590	0.9536	249	0.9356	0.9272
40	1.0070	1.0080	110	0.9826	0.9803	180	0.9587	0.9532	250	0.9352	0.9268
41	1.0067	1.0076	111	0.9823	0.9799	181	0.9584	0.9528	251	0.9349	0.9264
42	1.0063	1.0072	112	0.9819	0.9795	182	0.9580	0.9524	252	0.9346	0.9260
43	1.0060	1.0068	113	0.9816	0.9791	183	0.9577	0.9520	253	0.9342	0.9257
44	1.0056	1.0064	114	0.9813	0.9787	184	0.9574	0.9517	254	0.9339	0.9253
45	1.0053	1.0060	115	0.9809	0.9783	185	0.9570	0.9513	255	0.9336	0.9249
46	1.0049	1.0056	116	0.9806	0.9779	186	0.9567	0.9509	256	0.9332	0.9245
47	1.0046	1.0052	117	0.9802	0.9775	187	0.9563	0.9505	257	0.9329	0.9242
48	1.0042	1.0048	118	0.9799	0.9771	188	0.9560	0.9501	258	0.9326	0.9238
49	1.0038	1.0044	119	0.9795	0.9767	189	0.9557	0.9498	259	0.9322	0.9234
50	1.0035	1.0040	120	0.9792	0.9763	190	0.9553	0.9494	260	0.9319	0.9231
51	1.0031	1.0036	121	0.9788	0.9760	191	0.9550	0.9490	261	0.9316	0.9227
52	1.0028	1.0032	122	0.9785	0.9756	192	0.9547	0.9486	262	0.9312	0.9223
53	1.0024	1.0028	123	0.9782	0.9752	193	0.9543	0.9482	263	0.9309	0.9219
54	1.0021	1.0024	124	0.9778	0.9748	194	0.9540	0.9478	264	0.9306	0.9216
55	1.0017	1.0020	125	0.9775	0.9744	195	0.9536	0.9475	265	0.9302	0.9212
56	1.0014	1.0016	126	0.9771	0.9740	196	0.9533	0.9471	266	0.9299	0.9208
57	1.0010	1.0012	127	0.9768	0.9736	197	0.9530	0.9467	267	0.9296	0.9205
58	1.0007	1.0008	128	0.9764	0.9732	198	0.9526	0.9463	268	0.9293	0.9201
59	1.0003	1.0004	129	0.9761	0.9728	199	0.9523	0.9460	269	0.9289	0.9197
60	1.0000	1.0000	130	0.9758	0.9725	200	0.9520	0.9456	270	0.9286	0.9194
61	0.9997	0.9996	131	0.9754	0..9721	201	0.9516	0.9452	271	0.9196	0.9190
62	0.9993	0.9992	132	0.9751	0.9717	202	0.9513	0.9448	272	0.9279	0.9186
63	0.9990	0.9988	133	0.9747	0.9713	203	0.9509	0.9444	273	0.9276	0.9182
64	0.9986	0.9984	134	0.9744	0.9709	204	0.9506	0.9441	274	0.9273	0.9179
65	0.9983	0.9980	135	0.9740	0.9705	205	0.9503	0.9437	275	0.9269	0.9175
66	0.9979	0.9976	136	0.9737	0.9701	206	0.9499	0.9433	276	0.9266	0.9171
67	0.9976	0.9972	137	0.9734	0.9697	207	0.9496	0.9429	277	0.9263	0.9168
68	0.9972	0.9968	138	0.9730	0.9693	208	0.9493	0.9425	278	0.9259	0.9164
69	0.9969	0.9964	139	0.9727	0.9690	209	0.9489	0.9422	279	0.9256	0.9160

Table 2.03 Temperature-Volume Corrections for Asphalt (continued)

Observed Temperature, °F	A	B	Observed Temperature, °F	A	B	Observed Temperature, °F	A	B	Observed Temperature, °F	A	B
280	0.9253	0.9157	335	0.9073	0.8956	390	0.8896	0.8760	445	0.8721	0.8567
281	0.9250	0.9153	336	0.9070	0.8952	391	0.8892	0.8756	446	0.8718	0.8564
282	0.9246	0.9149	337	0.9066	0.8949	392	0.8889	0.8753	447	0.8715	0.8560
283	0.9243	0.9146	338	0.9063	0.8945	393	0.8886	0.8749	448	0.8714	0.8557
284	0.9240	0.9142	339	0.9060	0.8942	394	0.8883	0.8746	449	0.8709	0.8554
285	0.9236	0.9138	340	0.9057	0.8938	395	0.8880	0.8742	450	0.8705	0.8550
286	0.9233	0.9135	341	0.9053	0.8934	396	0.8876	0.8738	451	0.8702	0.8547
287	0.9230	0.9131	342	0.9050	0.8931	397	0.8873	0.8735	452	0.8699	0.8543
288	0.9227	0.9127	343	0.9047	0.8927	398	0.8870	0.8731	453	0.8696	0.8540
289	0.9223	0.9124	344	0.9044	0.8924	399	0.8867	0.8728	454	0.8693	0.8536
290	0.9220	0.9120	345	0.9040	0.8920	400	0.8864	0.8724	455	0.8690	0.8533
291	0.9217	0.9116	346	0.9037	0.8917	401	0.8861	0.8721	456	0.8687	0.8529
292	0.9213	0.9-113	347	0.9034	0.8913	402	0.8857	0.8717	457	0.8683	0.8526
293	0.9210	0.9109	348	0.9031	0.8909	403	0.8854	0.8717	458	0.8680	0.8522
294	0.9207	0.9105	349	0.9028	0.8906	404	0.8851	0.8710	459	0.8677	0.8519
295	0.9204	0.9102	350	0.9024	0.8902	405	0.8848	0.8707	460	0.8674	0.8516
296	0.9200	0.9098	351	0.9021	0.8899	406	0.8845	0.8703	461	0.8671	0.8512
297	0.9197	0.9094	352	0.9018	0.8895	407	0.8841	0.8700	462	0.8668	0.8509
298	0.9194	0.9097	353	0.9015	0.8891	408	0.8838	0.8696	463	0.8665	0.8505
299	0.9190	0.9087	354	0.9011	0.8888	409	0.8835	0.8693	464	0.8661	0.8502
300	0.9187	0.9083	355	0.9008	0.8884	410	0.8832	0.8689	465	0.8658	0.8498
301	0.9186	0.9080	356	0.9005	0.8881	411	0.8829	0.8686	466	0.8655	0.8495
302	0.9181	0.9076	357	0.9002	0.8877	412	0.8826	0.8682	467	0.8652	0.8492
303	0.9177	0.9072	358	0.8998	0.8873	413	0.8822	0.8679	468	0.8649	0.8488
304	0.9174	0.9069	359	0.8995	0.8870	414	0.8819	0.8675	469	0.8646	0.8485
305	0.9171	0.9065	360	0.8992	0.8866	415	0.8816	0.8672	470	0.8643	0.8481
306	0.9167	0.9061	361	0.8989	0.8863	416	0.8813	0.8668	471	0.8640	0.8478
307	0.9164	0.9058	362	0.8986	0.8859	417	0.8810	0.8665	472	0.8636	0.8474
308	0.9161	0.9054	363	0.8982	0.8856	418	0.8806	0.8661	473	0.8633	0.8471
309	0.9158	0.9050	364	0.8979	0.8852	419	0.8803	0.8658	474	0.8630	0.8468
310	0.9154	0.9047	365	0.8976	0.8848	420	0.8800	0.8654	475	0.8627	0.8464
311	0.9151	0.9043	366	0.8973	0.8845	421	0.8797	0.8651	476	0.8624	0.8461
312	0.9148	0.9039	367	0.8949	0.8841	422	0.8794	0.8647	477	0.8621	0.8457
313	0.9145	0.9036	389	0.8966	0.8838	423	0.8791	0.8644	478	0.8618	0.8454
314	0.9141	0.9032	369	0.8963	0.8834	424	0.8787	0.8640	479	0.8615	0.8451
315	0.9138	0.9028	370	0.8960	0.8831	425	0.8784	0.8637	480	0.8611	0.8447
316	0.9135	0.9025	371	0.8957	0.8827	426	0.8781	0.8633	481	0.8608	0.8444
317	0.9132	0.9021	372	0.8953	0.8823	427	0.8778	0.8630	482	0.8605	0.8440
318	0.9128	0.9018	373	0.8950	0.8820	428	0.8775	0.8626	483	0.8602	0.8437
319	0.9125	0.9014	374	0.8947	0.8816	429	0.8772	0.8623	484	0.8599	0.8433
320	0.9122	0.9010	375	0.8944	0.8813	430	0.8768	0.8619	485	0.8596	0.8430
321	0.9118	0.9007	376	0.8941	0.8809	431	0.8765	0.8616	486	0.8593	0.8427
322	0.9115	0.9003	377	0.8937	0.8806	432	0.8762	0.8612	487	0.8590	0.8423
323	0.9112	0.9000	378	0.8934	0.8802	433	0.8759	0.8609	488	0.8587	0.8420
324	0.9109	0.8996	379	0.8931	0.8799	434	0.8756	0.8605	489	0.8583	0.8416
325	0.9105	0.8992	380	0.8928	0.8795	435	0.8753	0.8602	490	0.8580	0.8413
326	0.9102	0.8989	381	0.8924	0.8792	436	0.8749	0.8599	491	0.8577	0.8410
327	0.9099	0.8985	382	0.8921	0.8988	437	0.8746	0.8595	492	0.8574	0.8406
328	0.9096	0.8981	383	0.8918	0.8784	438	0.8743	0.8592	493	0.8571	0.8403
329	0.9092	0.8978	384	0.8915	0.8781	439	0.8740	0.8588	494	0.8568	0.8399
330	0.9089	0.8974	385	0.8912	0.8777	440	0.8737	0.8585	495	0.8565	0.8396
331	0.9086	0.8971	386	0.8908	0.8774	441	0.8734	0.8581	496	0.8562	0.8383
332	0.9083	0.8967	387	0.8905	0.8770	442	0.8731	0.8578	497	0.8559	0.8389
333	0.9079	0.8963	388	0.8902	0.8767	443	0.8727	0.8574	498	0.8556	0.8386
334	0.9076	0.8960	389	0.8899	0.9763	444	0.8724	0.8571	499	0.8558	0.8383
									500	0.8549	0.8379

[1] Use column A for factors asphalts with API gravity at 60°F of 14.9° or less or with specific gravity 60/60°F of 0.967 or higher. Use column B factors for asphalts with API gravity at 60°F from 15.0° to 34.9° or with specific gravity 60/60°F from 0.850 to 0.966.

Table 2.04 Aggregate Classifications

Class	Type	Family
Sedimentary	Calcareous	Limestone Dolomite
	Siliceous	Shale Sandstone Chert Conglomerate[1] Breccia[1]
Metamorphic	Foliated	Gneiss Schist Amphibolite Slate
	Nonfoliated	Quartzite Marble Serpentinite
Igneous	Intrusive (coarse-grained)	Granite[2] Syenite[2] Diorite[2] Gabbro Periodotite Pyroxenite Hornblendite
	Extrusive (fine-grained)	Obsidian Pumice Tuff Rhyolite[2,3] Trachyte[2,3] Andesite[2,3] Basalt[2] Diabase

[1] May also be composed partially or entirely of calcareous materials.
[2] Frequently occurs as a porphyritic rock.
[3] Included in general term "felsite" when constituent minerals cannot be determined quantitatively.

- End products of chemical action or evaporation (salt, gypsum).
- Combinations of these types of materials.

Two terms often applied to sedimentary rocks are *siliceous* and *calcareous*. Siliceous sedimentary rocks are those which contain a high percentage of silica. Rocks containing a high percentage of calcium carbonate (limestone) are called calcareous.

Igneous Igneous rocks consist of molten material (magma) that has cooled and solidified. There are two types of igneous rock: *extrusive* and *intrusive*.

Extrusive igneous rock is formed from material that has poured out onto the earth's surface during a volcanic eruption or similar geologic activity. Because exposure to the atmosphere allows the material to cool quickly, the resulting rock has a glass-like appearance and structure. Rhyolite, andesite and basalt are examples of extrusive rock.

Intrusive rock forms from magma trapped deep within the earth's crust. Trapped in the earth, the magma cools and hardens slowly, allowing a crystalline structure to form. Examples of igneous rock are: granite, diorite and gabro. Earth movement and erosion processes bring intrusive rock to the earth's surface where it is quarried and used.

Metamorphic Metamorphic rock is generally sedimentary or igneous rock that has been changed by intense pressure and heat within the earth. Because such formation processes are complex, it is often difficult to determine the exact origin of a particular metamorphic rock.

Many types of metamorphic rock have a distinct characteristic feature: the minerals are arranged in parallel planes or layers. Splitting the rock along its planes is much easier than splitting it in other directions. Metamorphic rock that exhibits this type of structure is termed *foliated*. Examples of foliated rock are gneisses, schists (formed from igneous material) and slate (formed from shale, a sedimentary rock).

Not all metamorphic rock is foliated. Marble (formed from limestone) and quartzite (formed from sandstone) are common types of metamorphic rock without foliation.

▶▶Aggregate Sources

Aggregates for HMA are generally classified according to their sources. They include natural aggregates, processed aggregates and synthetic or artificial aggregates.

Natural Aggregates Natural aggregates are those used with little or no processing. They are made up of particles produced by natural erosion and degradation, such as the action of wind, water, moving ice and chemicals. The shape of individual particles is largely a result of erosion. Glaciers, for example, often produce rounded boulders and pebbles. Similarly, flowing water produces smoothly rounded particles.

The two major types of natural aggregates used in pavement construction are *gravel* and *sand*. Gravel is usually defined as particles 6.35 mm (1/4 in.) or larger in size. Sand is defined as particles smaller than 6.35 mm but larger than .075 mm (No. 200). Particles smaller than .075 mm are considered mineral filler.

Gravels and sands are further classified by their source. Materials quarried from an open pit and used without further processing are referred to as pit-run materials. Similarly, materials taken from stream banks are referred to as bank-run materials.

Gravel deposits vary widely in composition, but usually contain some sand and silt. Sand deposits ordinarily contain some clay and silt. Beach sands (some of which are now far inland) consist of fairly uniform size particles, while river sand often contains a variety of gravel, silt and clay.

Processed Aggregates Processed aggregates have been quarried, crushed and/or screened in preparation for use. There are two basic sources of processed aggregates: natural gravels that are crushed to make them more suitable for use in HMA, and fragments of bedrock and large stones that must be reduced in size.

Rock is crushed for three reasons:

- To reduce the size and improve the distribution and range (gradation) of particle sizes.
- To change the surface texture of the particles from smooth to rough.
- To change particle shape from round to angular.

Screening the materials after crushing classifies the particles into specific gradation ranges. Maintaining specific aggregate gradation is a critical element in producing quality HMA. Proper control of the crushing operation determines whether the resulting aggregate gradation meets job requirements.

Crushing some types of rocks, such as limestone, produces substantial quantities of smaller particles. In most operations these fractions are separated from all particles 6.35 mm (1/4 in.) in diameter or larger, and are used as crushed sand or processed further to a maximum particle size of 0.60 mm (No. 30).

Synthetic Aggregates Synthetic or artificial aggregates are the product of chemical or physical processing of materials. Some are by-products of industrial production processes such as ore refining. Others are produced specifically for use as aggregate by processing raw materials.

Blast-furnace slag is the most commonly used by-product aggregate. It is a nonmetallic substance that rises to the surface of molten iron during the smelting process. When drawn off the surface of the iron, the slag is reduced into small particles, either by quenching it immediately in water or crushing it after is has cooled.

Manufactured synthetic aggregates are relatively new in the HMA industry. They are manufactured by firing clay, shale, processed diatomaceous earth, volcanic glass, slag and other materials. The end products are typically lightweight and have unusually high resistance to wear. Synthetic aggregates have been used in bridge-deck and roof-deck paving, as well as pavement surface layers where maximum skid resistance is required.

►► Aggregate Production, Stockpiling, Handling and Sampling

Most contracting agencies do not specify handling and stockpiling procedures. Instead, they specify aggregate gradations and quality requirements. However, handling and stockpiling practices affect aggregate suitability. Therefore, sampling and testing are performed to verify that specifications are met. Correct sampling and testing procedures must be followed to ensure the samples selected are representative and are tested properly.

Production of Aggregates Special care must be taken removing soil overburden. This is particularly important where the overburden contains clay, vegetation or other materials that can adversely affect pavement performance. Some overburden material may provide an acceptable filler; however, it should be removed and processed separately.

Operations in pits and quarries must often work around clay lenses, shale seams and other deposits of unsuitable material embedded in the rock formation. To avoid contamination and ensure uniform aggregate gradation, excavation may have to be done along a horizontal bench, or from top to bottom of the formation's vertical face.

It is essential to thoroughly evaluate aggregates after crushing and screening to ensure they meet quality and gradation requirements. At commercial production facilities, where aggregate production can be continuous throughout the paving season, one or two quality evaluations per season may be satisfactory. Where an operation is starting up for the first time, evaluations of aggregate prior to their use in paving mixtures should be done regularly.

Stockpiling Good stockpiling procedures are crucial to HMA production. Properly stockpiled aggregates retain their gradation. Poorly stockpiled aggregates segregate (separate by size) and gradation varies within the stockpile.

Clean, well drained surfaces should be used for stockpiling aggregates. Precautions should be taken to keep stockpiles separated to maintain proper gradation. This is achieved by keeping stockpiles widely spaced; by using bulkheads between stockpiles; or by storing aggregate in bins. Bulkheads should extend to the full depth of the stockpiles.

Sands, crushed fine aggregate, and aggregates consisting of a single-size particle (especially small particles) can be stockpiled by almost any method with very little segregation. However, materials containing a range of particle sizes require certain stockpiling precautions. Segregation of graded aggregates can be minimized if coarse and fine material are separated at the site and blended in proper proportion prior to the mixing operation.

If not separated, certain stockpiling guidelines should be followed. The first guideline is to control the shape of the stockpile. When aggregate containing both coarse and fine materials is heaped into a stockpile with sloped sides, the coarse particles tend to roll down the slope and accumulate at the bottom.

The best method of stockpiling aggregates consisting of a range of different-size particles is to build the stockpile in layers. Such layers minimize segregation caused by gravity. If the aggregate is delivered by truck, loads should be placed close together over the surface of the stockpile. The volume of each truckload determines the thickness of each layer. When a crane is used to stockpile aggregate, each bucketload should be carefully placed to ensure uniform layer thickness.

Mineral fillers are usually stored in bins, silos or bags to prevent them from blowing away and being exposed to moisture. More details on aggregate stockpiling are provided in Chapter 4.

Handling All handling degrades individual aggregate particles to some extent. This can cause particle segregation where different-size aggregate particles are involved. Therefore, handling should be minimized to prevent degradation and segregation that could make the aggregate unsuitable for use.

Necessary handling includes removing aggregate from stockpiles for further processing or mixing in the hot mix facility. There are no set rules for this operation, but one general guideline is usually applicable: use a front-end loader or clamshell to remove material from a near-vertical face of the stockpile. Having a bulldozer or other tracked vehicle working on top of the stockpile increases the probability of serious degradation.

Sampling During the process of producing, stockpiling and handling aggregates, good quality control procedures:

- Ensure that only satisfactory material is used in the HMA.
- Provide a permanent record as evidence that the materials meet job specifications.

Obviously, it is not practical to test all the aggregate being produced or to test all the contents of a stockpile. It is feasible only to test samples of these materials. For test results to be accurate, the selected sample must be truly representative of the stockpile. Proper sampling techniques are very important (see Chapter 4).

►►Aggregate Properties and Evaluation

Aggregate makes up 90-95 percent by weight of HMA. This makes the quality of the aggregate a critical factor in pavement performance. In addition to quality, there are other criteria that influence aggregate selection for a particular project such as cost and availability. An aggregate that meets cost and availability requirements must still have certain properties to be considered suitable for use in quality HMA. These properties include:

- Maximum particle size
- Specific Gravity
- Toughness
- Absorption

- Aggregate gradation
- Cleanliness, or clay content
- Particle shape
- Moisture susceptibility

Another characteristic of aggregate that affects mixture behavior is surface texture. However, since there is no standard method for evaluating surface texture directly, it is not generally specified. A rough surface texture increases pavement strength because it prevents particles from moving easily past one another. A non-polishing surface texture will maintain a higher coefficient of friction and provide a more skid-resistant surface for safer traffic operation. In addition, asphalt films cling more readily to rough surfaces than to smooth ones.

Maximum Particle Size All HMA specifications require aggregate particles to be within a certain range of sizes and for each size of particle to be present in a certain proportion. This distribution of various particle sizes within the aggregate used is called the aggregate, or mix, gradation. To determine whether or not an aggregate gradation meets specifications requires an understanding of how particle size and gradation are measured.

Because specifications list a maximum particle size for each aggregate used, the size of the largest particles in the sample must be determined. There are two designations for maximum particle size:

- *Nominal maximum particle size* is designated as one sieve size larger than the first sieve to retain more than 10 percent in a standard series of sieves.
- *Maximum particle size* is defined as one sieve size larger than the nominal maximum particle size. Typically, this will be the smallest sieve through which 100 percent of the aggregate particles pass.

To illustrate the difference between the two designations, consider this example:

A sample of aggregate to be used in a paving mixture is put through a sieve analysis. All of the material passes through the 25 mm (1 in.) sieve and falls into the 19 mm (3/4 in.) sieve directly below. The 19 mm (3/4 in.) sieve retains 4 percent of the aggregate particles. The 12.5 mm (1/2 in.) sieve, directly below the 19 mm (3/4 in.) sieve, retains a total of 18 percent of the aggregate particles. In this case, the *nominal maximum* size is 19 mm (3/4 in.), and the *maximum* size is 25 mm (1 in.).

HMA is classified according to either its maximum size or its nominal maximum size (as in Superpave mixtures). Therefore, according to the maximum size of the aggregate described in the example, the mix would be termed a 25 mm (1 in.) mix. According to its nominal maximum size, the mixture would be a 19 mm (3/4 in.) mix.

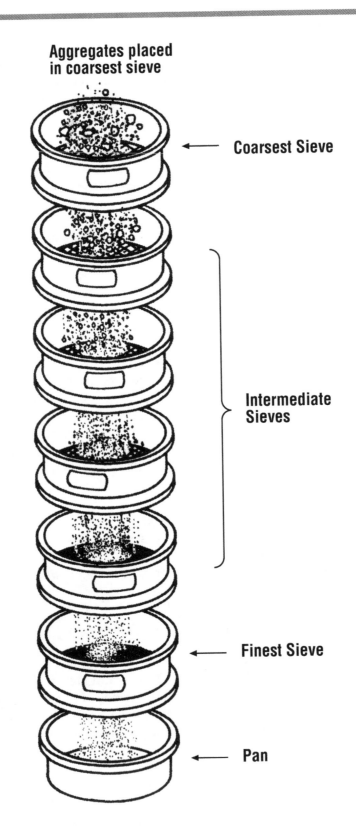

Aggregates placed in coarsest sieve

← **Coarsest Sieve**

Intermediate Sieves

← **Finest Sieve**

← **Pan**

Figure 2.18 Sieve Analysis

Table 2.05 Typical Composition of Hot Mix Asphalt

Sieve Size	Mix Designation and Nominal Maximum Size of Aggregate				
	37.5 mm (1½ in.)	25.0 mm (1 in.)	19.0 mm (¾ in.)	12.5 mm (½ in.)	9.5 mm (⅜ in.)
	Total Percent Passing (by weight)				
50 mm (2 in.)	100	–	–	–	–
37.5 mm (1½ in.)	90 to 100	100	–	–	–
25.0 mm (1 in.)	–	90 to 100	100	–	–
19.0 mm (¾ in.)	56 to 80	–	90 to 100	100	–
12.5 mm (½ in.)	–	56 to 80	–	90 to 100	100
9.5 mm (⅜ in.)	–	–	56 to 80	–	90 to 100
4.75 mm (No. 4)	23 to 53	29 to 59	35 to 65	44 to 74	55 to 85
2.36 mm (No. 8)*	15 to 41	19 to 45	23 to 49	28 to 58	32 to 67
1.18 mm (No. 16)	–	–	–	–	–
0.60 mm (No. 30)	–	–	–	–	–
0.30 mm (No. 50)	4 to 16	5 to 17	5 to 19	5 to 21	7 to 23
0.15 mm (No. 100)	–	–	–	–	–
0.075 mm (No. 200)**	0 to 5	1 to 7	2 to 8	2 to 10	2 to 10
Asphalt Binder, weight percent of Total Mixture†	3 to 8	3 to 9	4 to 10	4 to 11	5 to 12
	Suggested Coarse Aggregate Sizes				
	4 and 67 or 4 and 68	5 and 7 or 57	67 or 68 or 6 and 8	7 or 78	8

* In considering the total grading characteristics of an asphalt paving mixture, the amount passing the 2.36 mm (No. 8) sieve is a significant and convenient field control point between fine and coarse aggregate. Gradings approaching the maximum amount permitted to pass the 2.36 mm (No. 8) sieve will result in pavement surfaces having comparatively fine texture, while gradings approaching the minimum amount passing the 2.36 mm (No. 8) sieve will result in surfaces with comparatively coarse texture.

** The material passing the 0.075 mm (No. 200) sieve may consist of fine particles of the aggregates or mineral filler, or both. It shall be free from organic matter and clay particles and have a plasticity index not greater than 4 when tested in accordance with Method D 423 and Method D 424.

† The quantity of asphalt binder is given in terms of weight percent of the total mixture. The wide difference in the specific gravity of various aggregates, as well as a considerable difference in absorption, results in a comparatively wide range in the limiting amount of asphalt binder specified. The amount of asphalt required for a given mixture should be determined by appropriate laboratory testing or on the basis of past experience with similar mixtures, or by a combination of both.

Aggregate Gradation Particle gradation is determined by a sieve (or gradation) analysis of aggregate samples. A sieve analysis involves running the sample through a series of sieves, each of which has openings of specific sizes (Figure 2.18). Sieves are designated by the size of their openings. Coarse particles are retained on the upper sieves. Medium-size particles pass through to the mid-level sieves, and fine particles pass through to the lowest sieves.

The aggregate gradation is normally expressed as the percentage (by weight) of the total sample that passes through each sieve. It is determined by weighing the contents of each sieve

Table 2.06 Typical Sieve Sizes

Coarse Aggregate Sieve Designation		Fine Aggregate Sieve Designation	
Metric	U.S. Customary	Metric	U.S. Customary
63 mm	2½ in.	2.36 mm	No. 8
50 mm	2 in.	1.18 mm	No. 16
37.5 mm	1½ in.	0.60 mm	No. 30
25.0 mm	1 in.	0.30 mm	No. 50
19.0 mm	¾ in.	0.15 mm	No. 100
12.5 mm	½ in.	0.075 mm	No. 200
9.5 mm	⅜ in.		
4.75 mm	No. 4		

Table 2.07 Sieve Analysis Data Converted to Aggregate Gradation

Sieve Size	Retained Each Sieve (grams)	Passing Each Sieve (grams)	Total Percent Passing	Total Percent Retained	Passing-Retained,* Percent
9.0mm (¾ in.)	0	1135	100	0	5
12.5 mm (½ in.)	56	1079	95	5	15
9.5 mm (⅜ in.)	171	908	80	20	23
4.75 mm (No. 4)	262	646	57	43	18
2.36 mm (No. 8)*	203	443	39	61	16
0.60 mm (No. 30)	182	261	23	77	6
0.30 mm (No. 50)	68	193	17	83	5
0.15 mm (No. 100)	57	136	12	88	4.5
0.075 mm (No. 200)	51	85	7.5	92.5	7.5
Pan	85			100	

Total Weight = 1135 grams
* Passing designated sieve, retained on next smaller size.

following the sieve analysis and then calculating the percentage passing each sieve by one of several mathematical procedures. One method is to subtract the weight of the contents of each sieve from the weight of the material passing the previous sieve, resulting in the total weight passing each sieve. These weights are then converted to the percentage of the total sample passing each sieve.

HMA is graded by the percentages of different-size aggregate particles it contains. Table 2.05 illustrates five different HMA gradations. Sieves generally used in grading aggregate for HMA are shown in Table 2.06.

Certain terms are used in referring to aggregate fractions:

- *Coarse aggregate* – Material retained by the 2.36 mm (No. 8) sieve.
- *Fine aggregate* – Material passing the 2.36 mm (No. 8) sieve.

- *Mineral filler* – Fraction of fine aggregate that passes 0.60 mm (No. 30) sieve.
- *Mineral dust* – Fraction of fine aggregate passing the 0.075 mm (No. 200) sieve.

Mineral filler and mineral dust occur naturally with many aggregates and are produced as a by-product of crushing many types of rock. They are essential for producing a mixture that is dense, cohesive, durable, and resistant to water penetration. However, small changes in the amount or character of the mineral filler or dust can make significant changes in the quality and performance of HMA. Consequently, the type and amount of filler or dust used in any asphalt paving mixture must be carefully controlled.

The two methods for determining aggregate gradation are dry sieve analysis and wet sieve analysis. Dry sieve analysis alone is often used for coarser graded aggregate. When aggregate particles are coated with dust or silt-clay material, however, a washed sieve analysis should be performed.

Dry Sieve Analysis (ASTM C 136/AASHTO T 27)

- Sample for analysis is reduced by mechanical splitter or by quartering.
- Sample is dried to a constant weight.
- Sample is sieved into fractions.
- Weights of the fractions retained in each sieve and in the pan beneath the sieves are determined.

Washed-Sieve Analysis (ASTM C 117/AASHTO T 11)

- Sample for washed-sieve analysis is reduced, dried and weighed.
- Sample is then washed thoroughly to remove dust and silt-clay material (that which passes the .075 mm [No. 200] sieve).
- After washing, the sample is again dried and weighed. The difference between the weight before washing and the weight after washing determines the amount by weight of dust and silt-clay material in the original sample.
- Dry sieve analysis is performed on the washed sample (AASHTO T 27).

The method of determining the percentages of various-size particles from the weights of fractions obtained by sieve analysis is illustrated in Table 2.07. Gradations are usually expressed as a total percent passing each sieve (the total percent by weight of aggregate sample that passes through each sieve).

Graduations are sometimes expressed by two other methods:

- Total percent retained (the total percent by weight of aggregate sample retained by a given sieve).
- Total passing and retained (the total percent by weight of aggregate sample that passes through a given sieve, and is retained on the next smaller sieve).

After being calculated, aggregate gradation is often plotted as a grading curve. Two types of gradation charts are in general use: the semi-log chart (Figure 2.19) and the 0.45 power chart (Figure 2.20). The percent passing each sieve is recorded as a point on the appropriate vertical line. When one point is plotted for each sieve and its percent passing, the points are connected by a continuous line. The line represents the gradation curve of the aggregate analyzed.

Aggregate gradation specifications for a given job can also be presented graphically. On Figure 2.21, the specifications for the particular job are represented by the region between the

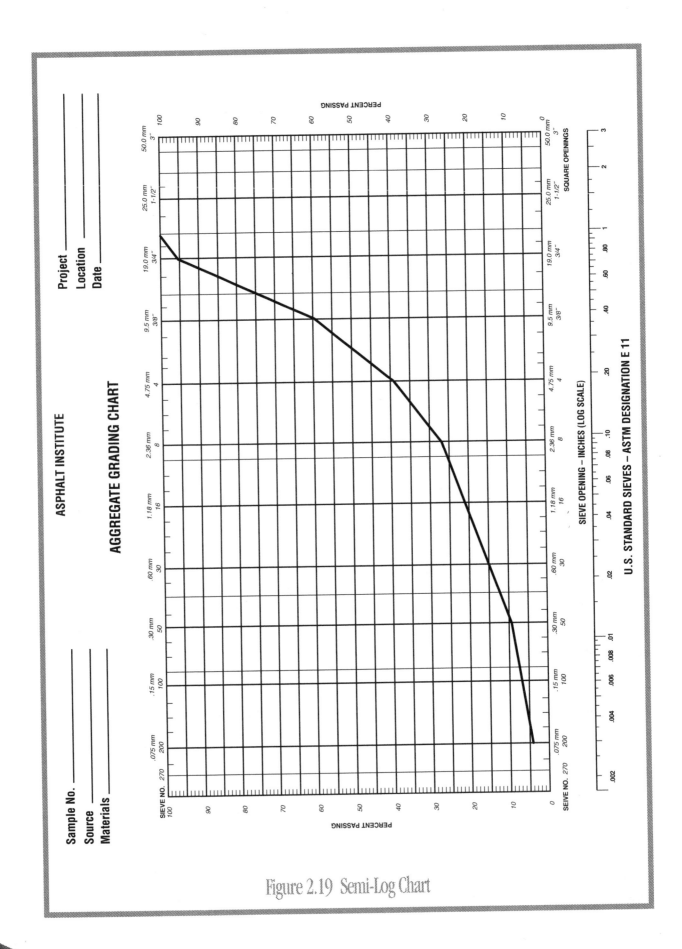

Figure 2.19 Semi-Log Chart

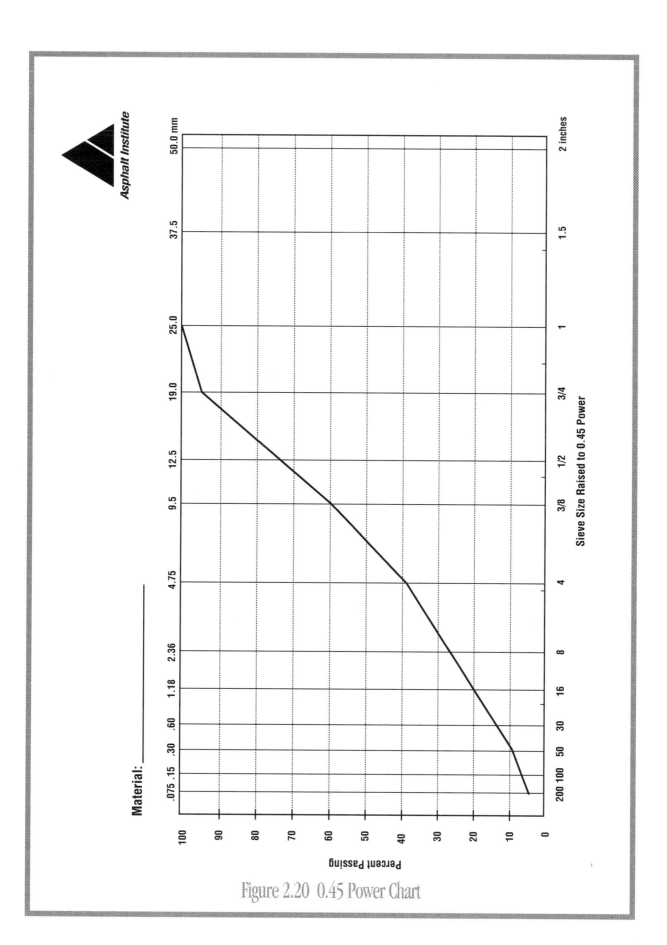

Figure 2.20 0.45 Power Chart

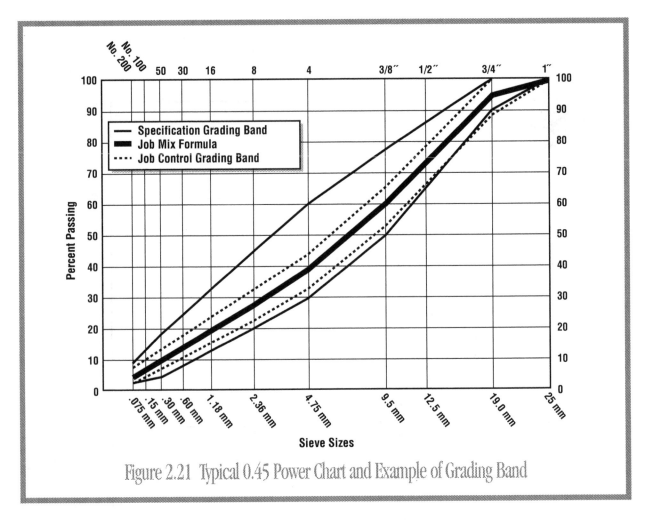

Figure 2.21 Typical 0.45 Power Chart and Example of Grading Band

thin solid lines. The paving mixture formula (job mix formula) is represented by the heavy solid line. The job control gradation band – established as the target for gradation control on the project – lies within the region bounded by the dotted lines. By plotting aggregate gradation along with the curves of the gradation specifications, one can tell immediately if the aggregate gradation falls within those specifications and targets.

Using Figure 2.21, we can examine what a gradation chart tells us. Taking the 9.5 mm (3/8-in.) sieve as an example, we see that the job control grading band permits between 54 percent and 66 percent of the aggregate to pass through. The job mix formula calls for 60 percent of the aggregate to pass through the 9.5 mm (3/8-in.) sieve. During mixing and construction, however, between 54 percent and 66 percent passing is the range used. A gradation chart helps to understand the gradations required by the specification gradation band, the job mix formula, and the job control gradation band.

The analysis of aggregate gradations and the combining of aggregates to obtain a desired gradation are important steps in HMA design. In combining aggregates, precise proportions of each must be determined to meet the target gradation. The aggregate gradation must meet the project specifications and yield a mix that meets the mix design criteria. The aggregate will normally be the most economical material available, provided that it complies with all of the quality requirements.

Specific Gravity The specific gravity of an aggregate is the ratio between the weight of a given volume of the aggregate and the weight of an equal volume of water. Specific gravity provides a means of expressing the weight-volume characteristics of materials. These characteristics are especially important in manufacturing pavement mixtures because the aggregate and asphalt in a mixture are proportioned by weight.

A ton of aggregate with a low specific gravity has a greater volume than a ton of aggregate with a higher specific gravity. Consequently, more asphalt must be added to a ton of aggregate with a low specific gravity in order to coat all the aggregate particles, than to a ton of aggregate with a high specific gravity.

Another critical reason for knowing the specific gravity of the aggregate is to aid in calculating the volumetric properties of the compacted mixtures. HMA must include a certain percentage of air spaces or voids (Chapter 3). These spaces perform important functions in the finished pavement. The percentage of air voids in compacted HMA specimens at various asphalt contents is calculated by a formula using the bulk specific gravities of the compacted mix and its maximum specific gravities at various asphalt contents. Maximum specific gravity is calculated using effective specific gravity of the aggregate along with the specific gravity of the asphalt.

All aggregates are porous to varying degrees. Because porosity affects the amount of asphalt needed to coat aggregate particles, as well as the percentage of air voids in the final mixture, three types of specific gravity measurements have been developed to take the porosity of an aggregate into consideration (Figure 2.22):

- Bulk specific gravity
- Apparent specific gravity
- Effective specific gravity

Bulk specific gravity is the specific gravity of a sample when the aggregate volume includes all pores in the sample.

Apparent specific gravity does not include pores and capillaries that would fill with water during soaking as part of the aggregate volume.

Effective specific gravity excludes from the aggregate's volume all pores and capillaries that absorb asphalt.

In an asphalt-aggregate mixture, the bulk specific gravity calculation of an aggregate would assume that no asphalt is absorbed into the aggregate pores. Calculating for apparent specific gravity would require an assumption that all of the pores are filled with asphalt. Except in rare cases, neither of these assumptions is true. Therefore, effective specific gravity, which takes into account the amount of asphalt absorbed into the pores, is the most nearly correct value for calculating the maximum specific gravity of the mix.

Cleanliness, or Clay Content Job specifications usually place a limit on the types and amounts of unsuitable material (vegetation, shale, soft particles, lumps of clay, etc.) permitted in the aggregate, particularly if the aggregate is known to contain quantities of such material. Excessive amounts of such material can have an adverse effect on pavement performance.

Aggregate cleanliness can be determined often by visual inspection, but a washed sieve analysis gives an accurate measurement of the percentage of unsuitable material finer than .075 mm (No. 200). The sand-equivalent test (ASTM D 2419/AASHTO T 176) is a method of determining the relative proportion of detrimental fine dust and clay-like material in the fraction (portion) of aggregate passing the 4.75 mm (No. 4) sieve.

Toughness Aggregates must be able to resist abrasion and degradation during manufacture, placing, compaction of the HMA and under actual traffic. Aggregates at or near the pavement

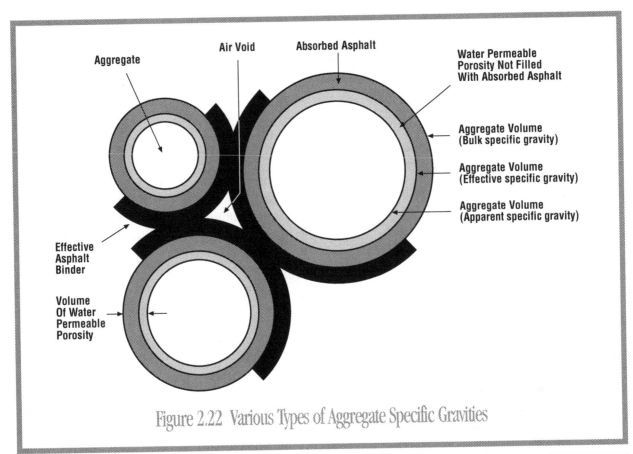

Aggregate

Air Void

Absorbed Asphalt

Water Permeable
Porosity Not Filled
With Absorbed Asphalt

Aggregate Volume
(Bulk specific gravity)

Aggregate Volume
(Effective specific gravity)

Aggregate Volume
(Apparent specific gravity)

Effective
Asphalt
Binder

Volume
Of Water
Permeable
Porosity

Figure 2.22 Various Types of Aggregate Specific Gravities

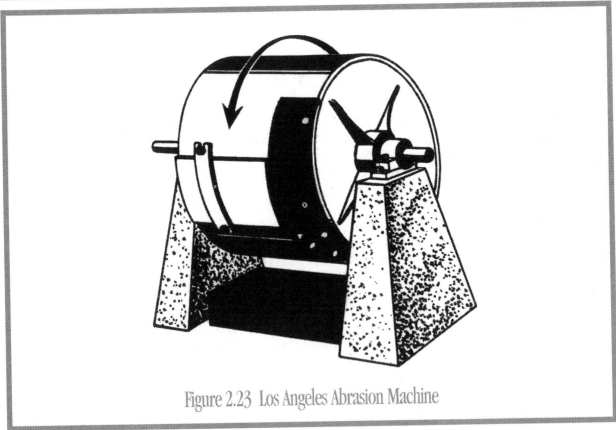

Figure 2.23 Los Angeles Abrasion Machine

Figure 2.24 Aggregate of Various Shapes and Surface Textures

surface must be tougher (more resistant to abrasion) than aggregates used in the lower layers of the pavement structure. This is because upper pavement layers receive more stress and wear from traffic loads.

The Los Angeles Abrasion Test (ASTM C 131 or C 535/AASHTO T 96) measures an aggregate's resistance to wear or abrasion. The test equipment is illustrated in Figure 2.23.

Particle Shape Particle shape (Figure 2.24) influences the workability of HMA during placement, as well as, the amount of force necessary to compact the mixture to the required density. During pavement life, particle shape influences the strength of the pavement structure. Aggregates at or near the surface must be of optimum shape.

Because angular particles tend to interlock when compacted, they usually resist displacement in the finished pavement. Particle interlocking is generally obtained with sharp-cornered, cubical-shaped particles. The term "crushed faces" has long been used to describe particle shape of coarse aggregates.

The Superpave asphalt mix design system adopted two procedures to quantify particle shape: coarse aggregate angularity (ASTM D 5821) and fine aggregate angularity (AASHTO T 304).

Absorption All aggregates are porous. Porosity determines how much liquid an aggregate particle absorbs when soaked.

The capacity of an aggregate to absorb water (or asphalt) is important information. If an aggregate is highly absorptive, it will continue to absorb asphalt after initial mixing at the plant, leaving less asphalt on its surface to bond aggregate particles together. Because of this, a porous aggregate requires significantly more asphalt to make a suitable mixture than a less porous aggregates does. Asphalt absorption is calculated during mix design (Chapter 3).

Highly porous, highly absorptive aggregates are not normally used unless they possess other characteristics that make them desirable. Blast furnace slag and other synthetic or manufactured aggregates are examples of highly porous, yet abrasion-resistant and skid-resistant material.

Moisture Susceptibility Stripping is a phenomenon in which the asphalt cement coating does not adhere to the aggregate after placement in the pavement structure. *Moisture susceptibility* is the term used to describe a mixture's tendency toward stripping. Why certain aggregates are moisture susceptible is not clearly understood.

Limestone, dolomite, and traprock have high affinities for asphalt and are also hydrophobic because they resist the efforts of water to strip asphalt from them. Hydrophilic aggregates have low affinities for asphalt. Consequently, they tend to separate from asphalt films when exposed to water. Siliceous aggregates (e.g., quartzite and some granites) are examples of aggregates that are prone to stripping and must be used cautiously.

Several test methods have been developed for determining moisture susceptibility. In one such test, the uncompacted aggregate-asphalt mixture is soaked in water and evaluated visually. In several other tests, two sets of mixture specimens are prepared and compacted. One is conditioned in water and the other is not. Both are then tested for indirect tensile strength. The ratio of the average strength of the conditioned samples to the unconditioned samples is considered to be an indicator of the aggregate's susceptibility to stripping. The Superpave asphalt mix design system adopted one of these strength comparison methods (AASHTO T 283) to evaluate HMA moisture susceptibility.

Introduction

In the production of hot mix asphalt (HMA), asphalt and aggregate are blended together in precise proportions. The relative proportions of these materials determine the physical properties of the mix and, ultimately, how the mix will perform as a finished pavement. There are three commonly used design procedures for determining suitable proportions of asphalt and aggregate in a mixture. They are the Marshall method, the Hveem method, and the Superpave system.

The Marshall and Hveem methods are both widely used for the design of HMA. The Superpave system is the newest system (discussed later in this chapter).

The objectives of this chapter are to:

- Know mix design characteristics and desirable properties of asphalt and asphalt mixtures.
- Recognize the causes of typical paving mixture deficiencies.
- Understand the relationship between mix design data and paving job specifications.
- Know the principal procedures involved in the Marshall and Hveem mix design methods and the Superpave mix design system.

Mixture Characteristics and Behavior

When a sample paving mixture is prepared in the laboratory, it can be analyzed to determine its probable performance in a pavement structure. The analysis focuses on four characteristics of the mixture and the influence those characteristics are likely to have on mix behavior. They are:

- Mix density
- Air voids
- Voids in the mineral aggregate
- Asphalt content

►►Mix Density

The density of the compacted mix is its unit weight or the weight of a specific volume of mix. Density is particularly important because high density of the finished pavement is essential for lasting pavement performance.

In mix design testing and analysis, density of the compacted specimen is usually expressed in kilograms per cubic meter (kg/m^3) or pounds per cubic foot (lbs/ft^3). It is calculated by multiplying the bulk specific gravity of the mix by the density of water [1,000 kg/m^3 (62.416 lbs/ft^3)]. The specimen density and the maximum theoretical density, both of which are determined in the laboratory, are each used as standards to determine if the density of the finished pavement meets specification requirements.

➤➤ Air Voids

Air voids are small pockets of air between the coated aggregate particles in the final compacted HMA. A certain percentage of air voids is necessary in the finished HMA to allow for a slight amount of compaction under traffic and a slight amount of asphalt expansion due to temperature increases. The allowable percentage of air voids in laboratory specimens is between 3 percent and 5 percent for surface and base courses, depending on the specific design.

The durability of an asphalt pavement is a function of the air void content. The lower the air voids, the less permeable the mixture becomes. An air void content that is too high provides passageways through the mix for the entrance of damaging air and water. An air void content that is too low can lead to flushing or bleeding.

Density and air void content are directly related. The higher the density is, the lower the voids in the mix will be, and vice versa. Job specifications usually require that pavement compaction achieve an air void content of less than 8 percent and more than 3 percent.

➤➤ Voids in the Mineral Aggregate

Voids in the mineral aggregate (VMA) are the intergranular void spaces that exist between the aggregate particles in a compacted paving mixture. VMA includes air voids and spaces filled with asphalt. VMA is a volumetric measurement expressed as a percentage of the total bulk volume of a compacted mix.

VMA (Figure 3.01) represents the space that is available to accommodate the effective volume of asphalt (i.e., all of the asphalt except the portion lost by absorption into the aggregate) and the volume of air voids necessary in the mixture. The more VMA in the dry aggregate, the more space is available for the films of asphalt. The durability of the mix increases with the film thickness on the aggregate particles. Therefore, specific minimum requirements for VMA are recommended and specified as a function of the aggregate size.

Minimum VMA is necessary to achieve an adequate asphalt film thickness, which results in a durable asphalt pavement. Increasing the density of the gradation of the aggregate to a point where below-minimum VMA values are obtained leads to thin films of asphalt and a low-durability mix. Therefore, economizing in asphalt content by lowering VMA is actually counter-productive and detrimental to pavement quality.

➤➤ Voids Filled with Asphalt

Voids filled with asphalt (VFA) is the percentage of intergranular void space between the aggregate particles (VMA) that contains or is filled with asphalt.

VFA is used to ensure that the effective asphalt part of the VMA in a mix is not too little (dry, poor durability) or too great (wet, unstable). The acceptable range of VFA varies depending upon the traffic level for the facility. Higher traffic requires a lower VFA, because mixture strength and stability is more of a concern. Lower traffic facilities require a higher range of VFA to increase HMA durability. A VFA that is too high, however, will generally yield a plastic mix.

➤➤ Asphalt Content

The asphalt content in the mixture is critical and must be accurately determined in the laboratory and then precisely controlled on the job. The asphalt content for a particular mix is established by using the criteria specific to the mix design method.

The optimum asphalt content of a mix is highly dependent on aggregate characteristics such as gradation and absorption. Aggregate gradation is directly related to optimum asphalt content.

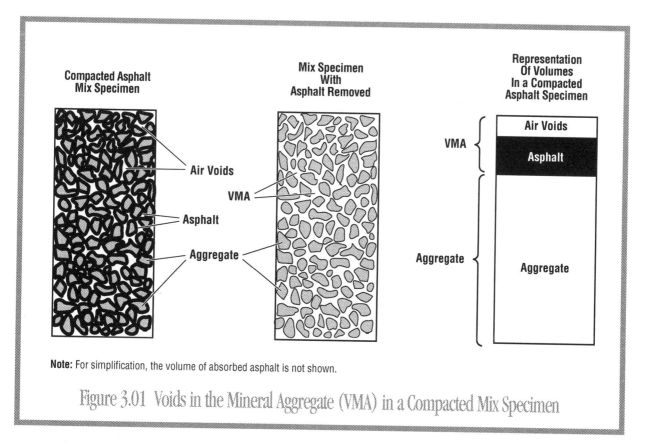

Note: For simplification, the volume of absorbed asphalt is not shown.

Figure 3.01 Voids in the Mineral Aggregate (VMA) in a Compacted Mix Specimen

The finer the mix gradation, the larger the total surface area of the aggregate and the greater the amount of asphalt required to uniformly coat the particles. Conversely, coarser mixes have less total aggregate surface area and demand less asphalt.

The relationship between aggregate surface area and optimum asphalt content is most pronounced when mineral filler containing high percentages of [particles passing the .075 mm (No. 200) sieve] is part of the mix. Variations in the amount and size of mineral filler at a constant asphalt content will cause changes in mix properties, resulting in mixes with too little or too much asphalt. Small increases in filler content will use up additional asphalt and produce a dry, unstable mix. Small decreases have the opposite effect. Too little filler results in a mixture that is too rich. If the filler has an abundance of very fine (less than 20 microns) particles, they will act as an asphalt extender, resulting in a mix that has too much asphalt.

If a mix contains too little or too much mineral filler, arbitrary adjustments are likely to have adverse effects. Instead, proper sampling and testing should be done to determine the cause of the variations and, if necessary, to establish a new mix design.

Aggregate absorption is critical in determining optimum asphalt content because enough asphalt must be added to the mix to allow for absorption and still coat the particles with an adequate film. The absorption for established aggregate sources is generally known, but requires careful testing when using new aggregate sources.

Two methods of expressing asphalt content are used in HMA technology: total asphalt content and effective asphalt content. Total asphalt content is the amount of asphalt that must be added to the mixture to produce the desired mix qualities. Effective asphalt content is based on the volume of asphalt not absorbed by the aggregate – of asphalt that effectively forms a bonding film on the aggregate surfaces. Effective asphalt content is calculated by subtracting the amount of absorbed asphalt, multiplied by the percentage of aggregate, from the total asphalt content.

Properties Considered in Mix Design

Hot mix asphalt pavements perform well because they are designed, produced and placed to provide certain desirable properties. These include stability, durability, impermeability, workability, flexibility, fatigue resistance, and skid resistance.

The final goal of mix design is to select a unique asphalt content that will achieve a balance among all of the desired properties. There is no single asphalt content that will maximize all of these properties. Instead, an asphalt content is selected on the basis of optimizing the properties necessary for the specific conditions.

➤➤ Stability

Stability of an asphalt pavement is its ability to resist shoving and rutting under traffic. A stable pavement maintains its shape and smoothness under repeated loading. An unstable pavement develops ruts, ripples (washboarding or corrugation) and other signs of mixture shifting.

Because stability specifications for a pavement depend on the traffic using the pavement, stability requirements can be established only after a thorough traffic analysis. Stability specifications should be high enough to handle traffic adequately, but not higher than traffic conditions require. A stability value that is too high produces a pavement that is too stiff and therefore less durable.

The stability of a mixture depends on internal friction and cohesion. Interparticle friction among the aggregate particles is related to aggregate characteristics such as shape and surface texture. Cohesion results from the bonding ability of the asphalt. A proper degree of both interparticle friction and cohesion in a mix prevents the aggregate particles from being moved past each other by the forces exerted by traffic. In general, more angular aggregate particles with rougher surface texture will increase the stability of the mix.

The binding force of cohesion increases with an increasing loading rate. Cohesion also increases as the viscosity of the asphalt increases, or as the pavement temperature decreases. Additionally, cohesion will increase with increasing asphalt content, up to a certain point. Past that point, increasing asphalt content creates too thick a film on the aggregate particles, resulting in loss of interparticle friction. Table 3.01 lists some of the causes and effects of insufficient stability in a pavement.

➤➤ Durability

The durability of an asphalt pavement is its ability to resist factors such as aging of the asphalt, disintegration of the aggregate, and stripping of the asphalt film from the aggregate. These factors result from weather, traffic, or a combination of the two.

Generally, the durability of a mixture can be enhanced by three methods:

- Designing the mix using a dense gradation of moisture-resistant aggregate
- Maximizing the asphalt film thickness on the aggregate
- Compacting the in-place mixture to 8 percent or less air voids

Asphalt film thickness is related to asphalt content and grade of asphalt binder. Thick asphalt films do not age and harden as rapidly as thin ones do. Consequently, the asphalt retains its original characteristics longer. Also, increased film thickness effectively seals off a greater percentage of interconnected air voids in the pavement, making it difficult for water and air to penetrate. A certain percentage of air voids must remain in the pavement to allow for expansion of the asphalt in hot weather and the pressure of traffic acting on the surface.

Table 3.01 Causes and Effects of Pavement Instability

Causes	Effects
Excess asphalt in mix	Washboarding, rutting and flushing or bleeding
Excess medium size sand in mixture	Tenderness during rolling and for period after construction, difficulty in compacting
Rounded aggregate, little or no crushed surfaces	Rutting and channeling

Table 3.02 Causes and Effects of Lack of Durability

Causes	Effects
Low asphalt content	Dryness or raveling
High void content through design or lack of compaction	Early hardening of asphalt followed by cracking or disintegration
Water susceptible (hydrophillic) aggregate in mixtures	Asphalt film strips from aggregate leaving an abraded, raveled or mushy pavement

A dense gradation of sound, tough, moisture-resistant aggregate contributes to pavement durability in three ways. It provides closer contact among aggregate particles, enhancing the impermeability of the mixture. A sound, tough aggregate resists disintegration under traffic loading. Moisture-resistant aggregate prevents the action of water and traffic from stripping off the asphalt film and leading to raveling of the pavement. Designing the mixture for maximum density and compacting the pavement to 8 percent or less air voids also minimizes the intrusion of air and water into the pavement.

A lack of sufficient durability in a pavement can have several causes and effects as shown in Table 3.02.

▶▶ Impermeability

Impermeability prevents the passage of air and water into or through the asphalt pavement. This characteristic is related to the void content of the compacted mixture, and much of the discussion on voids in mix design relates to impermeability. Even though void content is an indication of the potential for passage of air and water through a pavement, the character of these voids is more important than the number of voids. The size of the voids, whether or not they are interconnected, and the access of the voids to the surface of the pavement, all determine the degree of impermeability.

Although impermeability is important for durability of compacted paving mixtures, virtually all asphalt mixtures used in highway construction are permeable to some degree. This is acceptable as long as it is within specified limits. Causes and effects of poor impermeability values in dense-graded HMA are shown in Table 3.03.

Table 3.03 Causes and Effects of Permeability	
Causes	**Effects**
Low asphalt content	Thin asphalt films will cause early aging and raveling
High voids content in design mix	Water and air can easily enter pavement, causing oxidation and disintegration
Inadequate compaction	Will result in high voids in pavement, leading to water infiltration and low strength

▶▶ *Workability*

Workability describes the ease with which a paving mixture can be placed and compacted. Mixtures with good workability are relatively easy to place and compact; those with poor workability are difficult to place and compact. Changing mix design parameters, aggregate source, and/or gradation can improve workability.

Harsh mixtures (mixtures containing a high percentage of coarse aggregate) have a tendency to segregate during handling and also may be difficult to compact. Through the use of trial mixes in the laboratory, additional fine aggregate and perhaps asphalt, can be added to a harsh mix to make it more workable. Care should be taken to ensure that the altered mix meets all other design criteria.

Too high a filler content can also affect workability. It can cause the mix to become gummy, making it difficult to compact.

Workability is especially important where quite a bit of hand placement and raking (luting) around manhole covers, sharp curves, and other obstacles are required. It is important that mixtures used in such areas be highly workable.

Mixtures that can be too easily worked or shoved are referred to as tender mixes. Tender mixes are too unstable to place and compact properly. They are often caused by:

- A shortage of mineral filler.
- Too much medium-size sand.
- Smooth, rounded aggregate particles.
- And/or too much moisture in the mix.

Although not normally a major contributor to workability problems, asphalt does have some effect on workability. Because the temperature of the mix affects the viscosity of the asphalt, a temperature that is too low will make a mix unworkable, and a temperature that is too high may make it tender. Asphalt grade may also affect workability, as may the percentage of asphalt in the mix.

Table 3.04 lists some of the causes and effects related to workability of paving mixtures.

▶▶ *Flexibility*

Flexibility is the ability of an asphalt pavement to adjust to gradual settlements and movements in the subgrade without cracking. Since virtually all subgrades either settle (under loading) or rise (from soil expansion), flexibility is a desirable characteristic for all asphalt pavements.

Table 3.04 Causes and Effects of Workability Problems

Causes	Effects
Large maximum-size particle	Rough surface, difficult to place
Excessive coarse aggregate	May be hard to compact
A mix temperature that is too low	Uncoated aggregate, not durable, rough surface, hard to compact
Too much medium-size sand	Mix shoves under roller, remains tender
Low mineral filler content	Tender mix, highly permeable
High mineral filler content	Mix may be dry or gummy, hard to handle, not durable

Table 3.05 Causes and Effects of Poor Fatigue Resistance

Causes	Effects
Low asphalt content	Fatigue cracking
High design voids	Early aging of asphalt followed by fatigue cracking
Lack of compaction	Early aging of asphalt followed by fatigue cracking
Inadequate pavement thickness	Excessive bending followed by fatigue cracking

An open-graded mix or one with a high asphalt content is generally more flexible than a dense-graded mix or one with a low asphalt content. Sometimes the need for flexibility conflicts with the need for stability, so that trade-offs have to be made in selecting the optimum asphalt content.

➤➤ Fatigue Resistance

Fatigue resistance is the pavement's resistance to repeated bending under wheel loads (traffic). Research shows that air voids and asphalt viscosity have a significant effect on fatigue resistance. As the percentage of air voids in the pavement increases, either by design or lack of compaction, pavement fatigue resistance is drastically reduced. Likewise, a pavement containing asphalt that has aged and hardened significantly has reduced resistance to fatigue.

The thickness and strength characteristics of the pavement and the support of the subgrade also have a great deal to do with determining pavement life and preventing load-associated cracking. Thick, well-supported pavements do not bend as much under loading as thin or poorly supported pavements do. Therefore, they have longer fatigue lives. Table 3.05 presents a list of causes and effects of poor fatigue resistance.

Table 3.06 Causes and Effects of Poor Skid Resistance	
Causes	**Effects**
Excess asphalt	Bleeding, low skid resistance
Poorly textured or graded aggregate	Smooth pavement, potential for hydroplaning
Polishing aggregate in mixture	Low skid resistance

▶▶ Skid Resistance

Skid resistance is the ability of an asphalt surface to minimize skidding or slipping of vehicle tires, particularly when wet. For good skid resistance, tire tread must be able to maintain contact with the aggregate particles instead of riding on a film of water on the pavement surface (hydroplaning). Pavement skid resistance is typically measured at 65 km/hr (40 mi/hr) with a standard tread tire under controlled wetting of the pavement surface.

A rough pavement surface with many little peaks and valleys will have greater skid resistance than a smooth surface. Best skid resistance is obtained with rough-textured aggregate in a relatively open-graded mixture with an aggregate of about 9.5 mm (3/8 in.) to 12.5 mm (1/2 in.) maximum size. Besides having a rough surface, the aggregates must resist polishing (smoothing) under traffic. Calcareous aggregates polish more easily than siliceous aggregates. Unstable mixtures that tend to rut or bleed present serious skid resistance problems. A list of causes and effects relating to poor skid resistance is shown in Table 3.06.

Evaluation and Adjustment of Mix Design

When developing a specific mix design, several trial mixtures should be evaluated to determine which one best meets the design criteria. Each trial mix serves as a guide for evaluating and adjusting the trials that follow.

Initial trial mixes for establishing the job mix formula must have an aggregate gradation that is within job specifications. Where initial trial mixes fail to meet the design criteria, it is necessary to modify or redesign the mix using a different aggregate gradation or different aggregate sources.

Gradation curves (or lines) are helpful in making necessary adjustments in mix designs. Many designers find the 0.45 power chart convenient for adjusting aggregate gradations. For example, Figure 3.02 shows maximum density lines plotted on a 0.45 power gradation chart. The maximum density lines on a 0.45 power chart are determined by drawing a straight line from the origin at the lower left of the chart to the desired maximum particle size at the top.

The maximum density line represents the tightest, most dense gradation for an aggregate with a particular maximum size. Gradations that closely follow the maximum density line usually have low VMA values and must be adjusted away from it. Such adjustments increase VMA values, allowing the use of enough asphalt to obtain maximum durability without the mixture flushing.

Following are general guidelines for adjusting a trial mix in order to meet design criteria. The heading of each subsection describes the mixture condition needing correction. The suggestions outlined may not apply in all cases.

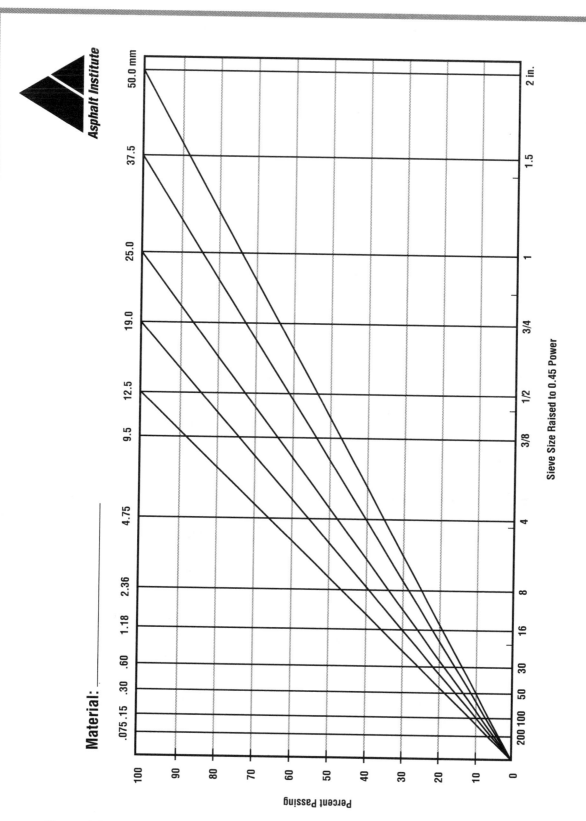

Figure 3.02 Maximum Density Curves Based on FHWA 0.45 Power Gradation Chart

➤➤ Voids Low, Stability Low

Voids may be increased in a number of ways. As a general approach to obtaining higher air voids and voids in the mineral aggregate, the aggregate gradation should be adjusted by adding more coarse or more fine aggregate.

Another way to increase air voids is to lower the asphalt content. This can be done only if excess asphalt is in the mix and the asphalt content is not reduced below the point where film thickness and, subsequently, pavement durability, are reduced below acceptable levels.

Increasing the amount of crushed materials and/or decreasing the amount of material passing the .075 mm (No. 200) sieve can improve stability and increase the air voids content. With some aggregates (quartz and similar rock types), however, the freshly fractured faces are as smooth as the water-worn faces, and an appreciable increase in stability is not achievable by crushing alone. By adding more manufactured sand, the void content can also be improved without sacrificing mix stability.

➤➤ Voids Low, Stability Satisfactory

Low void content may eventually result in instability due to plastic flow or flushing after the pavement has been exposed to traffic for a period of time because of particle reorientation and additional compaction. Insufficient void space may also result because of the higher amount of asphalt required to obtain high durability in finer mixes.

Degradation of a poor aggregate under the action of traffic may also subsequently lead to instability and flushing if the void content of the mix is not sufficient. For these reasons, mixes low in voids should be adjusted by one of the methods given above, even though the stability may initially appear satisfactory.

➤➤ Voids Satisfactory, Stability Low

Low stability when voids and aggregate gradation are satisfactory may indicate some deficiencies in the aggregate. Consideration should be given to improving the coarse aggregate particle shape by crushing or possibly increasing the maximum aggregate size. Aggregate particles with rougher texture and less rounded surfaces will exhibit more stability while maintaining or increasing the void content.

➤➤ Voids High, Stability Satisfactory

High voids are frequently associated with high permeability. High permeability, by permitting circulation of air and water through the HMA, may lead to premature hardening of the asphalt, raveling of the pavement surface, or possibly asphalt film stripping off the aggregate. Even though mix stability is satisfactory, the excessive void content should be reduced. Small reductions can be accomplished by increasing the mineral dust content of the mix. It may be necessary, however, to adjust the aggregate gradation closer to the maximum density line.

➤➤ Voids High, Stability Low

When the voids are high and the stability low, the void content should be reduced by the methods discussed above. If this adjustment does not improve both void content and stability, the type of aggregate used must be revised, as also described above.

Applications of Mix Design Testing

The importance of mix properties and mix design procedures is in the way they relate to construction control and inspection. The agency or authority responsible for paving construction usually establishes the mix design method and design requirements, which form an essential part of the construction specifications for asphalt pavements.

Mix design tests are both a means of establishing specifications and of checking that the paving mixture used on the roadway meets those specifications. Normally, mix design testing has four important applications in the overall construction:

- Preliminary design
- Source acceptance
- Quality control
- HMA compaction criteria

➤➤ Preliminary Design Testing

Preliminary design testing determines if prospective sources can provide aggregate that satisfies both gradation and mix design specifications. Results of preliminary design testing also indicate whether or not design requirements can be practically obtained within the framework of the specifications.

➤➤ Source Acceptance Testing

Source acceptance testing determines the blend of aggregates that will best satisfy both gradation and mix design requirements at an economical cost. This procedure can be accomplished by the agency upon notification of the materials' sources by the contractor, or it can be done initially by the contractor, with subsequent verification by the agency. The testing ensures the selection of proper materials and permits the contractor to begin stockpiling these materials at the HMA plant.

➤➤ Quality Control Testing

Quality control testing determines if the field produced HMA using the job mix formula meets the specification requirements. The job mix formula (JMF) is the "recipe" used by the plant to produce the paving mixture. The JMF includes the aggregate gradation and the selected asphalt content. Because variations in the mix are inevitable during production, the job mix formula has built-in tolerances that allow for reasonable variations in gradation and asphalt content.

Quality control tests are performed at the start of plant production to verify that the properties of the JMF can be met. Subsequent quality control testing involves periodically sampling the mixture produced by the plant and testing the properties of the samples. The results of the tests are compared with the JMF and the overall specification requirements. In those instances where irregularities occur and the limits of the JMF are exceeded, appropriate corrections will be required at the plant. Where the situation warrants, it may be necessary to reevaluate and redesign the paving mixture. Quality control and acceptance criteria are discussed in detail in Chapter 7.

➤➤ HMA Compaction Criteria

One of three methods is generally used to establish the minimum acceptable density for a compacted HMA pavement. Often termed the "target density," this requirement is determined by tests on mix produced at the plant at the time of placement. The three methods compare the in-place density to:

- The maximum theoretical density (or Rice density) of the mix
- The density of specimens compacted in the field laboratory
- The density of a control strip (test strip) on the paving site

Typical specifications require that each lot of the compacted mix have a density equal to or greater than 92 percent of the maximum theoretical density. When compared against field laboratory compacted specimens, the requirement is generally that the compacted pavement be at least 96 percent of the specimen density. When using a control strip, the specifications typically require that the density of the subsequently compacted pavement be anywhere from 98 percent to 100 percent of the density obtained in the control strip. See Chapter 7 for more information.

Objectives of Mix Design

The foregoing discussion provides a general view of the significance of mix design. The design of asphalt paving mixes is largely a matter of selecting and proportioning materials to obtain the desired properties in the finished construction. The overall objective of the design procedure is to determine an economical blend and gradation of aggregates (within the limits of the project specifications) and asphalt that yields a mix having:

- Sufficient asphalt to ensure a durable pavement
- Adequate mix stability to satisfy the demands of traffic without distortion or displacement
- Voids content high enough to allow for a slight amount of additional compaction under traffic loading without flushing, bleeding, and loss of stability, yet low enough to keep out harmful air and moisture
- Sufficient workability to permit efficient placement of the mix without segregation

The selected mix design is usually the one that best meets all of the established criteria. It should be accomplished with well-trained personnel using the proper materials and calibrated equipment and following the specified procedures. The mix design test procedures are also used for quality control testing of the produced mix, for HMA acceptance testing and for determining density of the completed pavement.

Marshall Mix Design Method

The Marshall method of designing paving mixtures was developed by Bruce Marshall, formerly Bituminous Engineer with the Mississippi State Highway Department. The Marshall test in its present form originated from an investigation started by the U.S. Army Corps of Engineers in 1943. Various methods for the design and control of asphalt mixtures were compared and evaluated to develop a simple method of asphalt pavement mixture design and control.

Since the Marshall method used portable equipment, the Corps adopted the Marshall method and developed and adapted it for both design and control of HMA in the field. Through extensive laboratory research, traffic tests and correlation studies, the Corps of Engineers improved and added certain features to Marshall's test procedure and ultimately developed mix design criteria.

The Marshall method is applicable only to hot mix asphalt using penetration, viscosity or PG graded asphalt cements and containing aggregates with maximum sizes of 25.0 mm (1 in.) or less. The method may be used for both laboratory design and field control of HMA.

The purpose of the Marshall method is to determine the optimum asphalt content for a particular blend of aggregate. The method also provides information about the properties of the resulting HMA, including density and void content, which are used during pavement construction.

The Marshall method uses standard test specimens 64 mm (2.5 in.) high and 102 mm (4 in.) in diameter. A series of specimens, each containing the same aggregate blend but varying in asphalt content, is prepared using a specific procedure to heat, mix and compact the asphalt aggregate mixtures.

The two principal features of the Marshall method of mix design are a density-voids analysis and a stability-flow test of the compacted test specimens.

►►Conducting Marshall Mix Design

This is a general description of the procedures followed in Marshall mix design. The complete and detailed procedure that must be followed is contained in ASTM D 1559 or AASHTO T 245. The Asphalt Institute publication *Mix Design Methods for Asphalt Concrete and other Hot Mix Types*, Manual Series No. 2 (MS-2), details the procedures and equipment for conducting Marshall mix design.

As discussed in Chapter 2, different aggregates and asphalts have different characteristics. These characteristics have a direct impact on the nature of the pavement itself. The first step in the design method is to determine what qualities (stability, durability, workability, skid-resistance, etc.) the paving mixture must have. The second step is to select a type of aggregate and a compatible type of asphalt that will combine to produce those qualities. Once this is done, test preparations can begin.

►►Selection of Material Samples

The first step is to gather samples of the asphalt and the aggregate that will be used in the actual paving mixture. It is important that the asphalt samples have characteristics identical to those of the asphalt that will be used in the final HMA. The same is true of the aggregate samples. The reason is simple: data derived from the mix design procedures determine the job mix formula (JMF) for the actual paving mixture. The JMF can be accurate only if the ingredients tested in the laboratory have characteristics identical to those of the ingredients used in the final product.

A wide variety of serious problems, ranging from poor workability of the mix to premature failure of the pavement itself, are historical results of variances between materials tested and materials actually used.

Figure 3.03 Determining Specific Gravity of an Aggregate Sample

Figure 3.04 Preparing Test Specimen for Marshall Mold

➤➤ Aggregate Preparation

Washed-Sieve Analysis The washed-sieve analysis is performed because it is important that all of the mineral dust in the aggregate be included in the test result. The washed-sieve analysis involves these steps:

1. Each aggregate sample is dried and weighed.
2. Each sample is washed thoroughly over a .075 mm (No. 200) sieve to remove any mineral dust coating the aggregate. The sieve retains all material larger than .075 mm.
3. The washed samples are again thoroughly dried.
4. The dry weight of each sample is recorded. By comparing the sample weight taken before washing to the sample weight taken after, the amount of mineral dust in each sample can be determined.

Specific Gravity The specific gravity of any substance is the weight-volume ratio of a unit of that substance compared to the weight-volume ratio of an equal unit of water. The specific gravity of an aggregate sample is determined by comparing the weight of a given volume of the aggregate to the weight of an equal volume of water at the same temperature (Figure 3.03). The specific gravity is expressed in multiples of the specific gravity of water (which is always equal to 1). For example, an aggregate sample that weighs two-and-one-half times as much as an equal volume of water has a specific gravity of 2.5.

The test procedures and method of calculating the specific gravity of the aggregate are given in ASTM C 127 and C 128 and AASHTO T 84 and T 85. This establishes a reference point for later specific gravity measurements taken to determine the proportions of aggregate, asphalt, and air voids in the test mixes.

➤➤ Preparing Mixture Test Samples

Test samples of possible paving mixtures are prepared, each containing a slightly different amount of asphalt. The range of asphalt contents used in the test samples is determined from previous experience with the aggregates in the mix. The range gives the laboratory a starting point for determining an exact asphalt content for the final mixture. The aggregate portion of the mixtures is formulated from the results of the sieve analysis.

The sample preparation procedures are:

1. Asphalt and aggregate are heated and thoroughly mixed together until all aggregate particles are coated. This simulates the heating and mixing processes at a HMA plant.
2. The hot asphalt mixtures are placed in preheated Marshall molds (Figure 3.04) in preparation for compaction by the Marshall drop hammer, which is also heated so it does not cool the mix surface when it strikes the mix.
3. The specimens are compacted by blows from the Marshall hammer (Figure 3.05). The number of blows (35, 50 or 75) depends on the design traffic for the mix. Both sides of each specimen receive the same number of compacting blows. After compaction is completed the specimens are cooled and then removed from the molds.

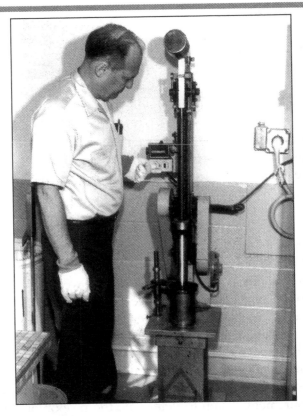

Figure 3.05 Marshall Drop Hammer Compacting Mix Specimen

➤➤ Marshall Test Procedures

There are three main test procedures in the Marshall test method: determination of bulk specific gravity, measurement of Marshall stability and flow, and analysis of specimen density and voids content.

Bulk Specific Gravity Determination As soon as the freshly compacted specimens have cooled to room temperature, the bulk specific gravity (ASTM D 2726 or D 1188/AASHTO T 166) of each specimen is determined. This measurement is essential for an accurate density/air voids analysis.

Stability and Flow Tests After the bulk specific gravities have been determined, the stability and flow tests are performed. Stability testing aims at measuring the mix's resistance to deformation under loads. Flow testing measures the amount of deformation that occurs in the mix under loading. The test procedure is:

1. The specimens are heated in a water bath to 60°C (140°F), representing the warmest in-service temperature that the pavement will normally experience.
2. The specimen is removed from the water bath, damp dried, and quickly placed in the Marshall apparatus (Figure 3.06). The apparatus consists of a device for exerting a load on the specimen and instruments for measuring the load and the flow. Each test result is normally plotted on a piece of graph paper.

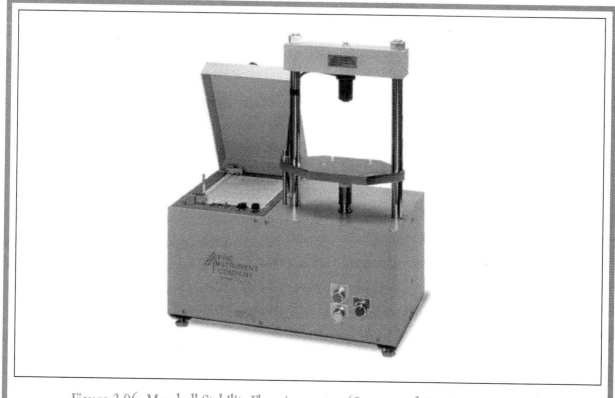

Figure 3.06 Marshall Stability-Flow Apparatus (Courtesy of Pine Instrument Co.)

3. The testing load is applied to the specimen at a constant rate of 51 mm (2 in.) per minute until failure occurs. Failure is defined as the maximum load the specimen will withstand.
4. The load at failure is recorded as the Marshall stability value and the amount of deformation is recorded as the flow value.

The Marshall stability value is the maximum load before the specimen yields or fails. Because Marshall stability indicates the resistance of the mix to deformation, there is a natural tendency to think that if a certain stability value is good, then a much higher value must be better. This tendency, however, will lead to the use of mixes with stability values that are too high.

For many engineering materials, the strength of the material frequently is a measure of its quality. However, this is not necessarily the case with HMA. Extremely high stability often is obtained at the expense of durability.

Marshall flow, measured in units equal to 0.25 mm (one-hundredth of an inch), represents the total deformation of the specimen. The deformation is a measure of the decrease in the vertical diameter of the specimen.

Mixes that have very low flow values and abnormally high Marshall stability values are considered too brittle and rigid for pavement service. Those with high flow values are considered too plastic and have a tendency to distort easily under traffic loads.

Density and Voids Analysis   Upon completion of the physical tests, a density and voids analysis is performed for each series of test specimens. The purpose of the analysis is to determine by calculation the percentage of air voids in the compacted mix.

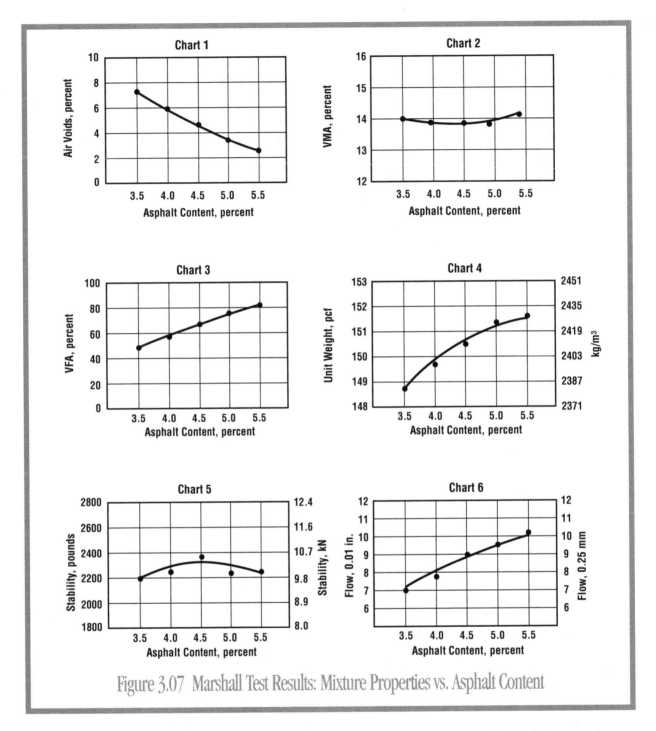

Figure 3.07 Marshall Test Results: Mixture Properties vs. Asphalt Content

- *Voids Analysis* – The air voids are the small pockets of air between the asphalt-coated aggregate particles. The percent air voids are calculated from the bulk specific gravity of each compacted specimen and the maximum specific gravity of the paving mixture (zero voids). The latter is determined directly by a standard test (ASTM D 2041/AASHTO T 209) on an uncompacted mix sample; or it can be calculated from the specific gravities of the asphalt and aggregate in the mix, with an appropriate allowance made for the amount of asphalt absorbed by the aggregate. The bulk specific gravities of compacted specimens had previously been determined by weighing specimens dry in air and saturated and immersed in water.

- *Density Analysis* – The average density or unit weight for each sample is determined by multiplying the bulk specific gravity of the mix by 1000 kg/m³ (62.4 lbs/ft³).

- *VMA Analysis* – The voids in the mineral aggregate (VMA) are defined as the intergranular void space between the aggregate particles in a compacted paving mixture that includes the air voids and the effective asphalt content expressed as a percent of the total volume. The VMA is calculated on the basis of the bulk specific gravity of the aggregate and is expressed as a percentage of the bulk volume of the compacted paving mixture. Therefore the VMA can be calculated by subtracting the volume of the aggregate, determined by its bulk specific gravity, from the bulk volume of the compacted paving mixture.

- *VFA Analysis* – The voids filled with asphalt (VFA) is the percentage of the intergranular void space between the aggregate particles (VMA) that is filled with asphalt. VMA includes asphalt and air. Therefore, the VFA is calculated by subtracting the air voids from the VMA and dividing by the VMA, and expressing the value as a percentage.

Analyzing Marshall Test Results

➤➤ Plotting Test Results

Plotting the Marshall test results on charts or graphs helps one understand the characteristics of each specimen in a test series. By studying the charts, you can determine the asphalt content that best meets all the mix design criteria. This is called the optimum asphalt content. The proportions of asphalt and aggregate in the final mixture are based on the optimum asphalt content. Figure 3.07 shows six Marshall test result charts. On each chart a "best fit" curve is used to represent the test values. The charts include plots of asphalt content versus:

- Air voids
- Voids in the mineral aggregate (VMA)
- Voids filled with asphalt (VFA)
- Unit weight (density)
- Marshall stability
- Marshall flow

➤➤ Trends in Test Data

When plotted on charts, the test results usually reveal reasonably consistent trends in the relationship between asphalt content and mix properties. Below are listed trends that can be seen by studying individual charts in Figure 3.07.

- The percentage of air voids decreases with increasing asphalt content (Chart 1).
- The percentage of voids in the mineral aggregate generally decreases slightly to a minimum value then increases with increasing asphalt contents (Chart 2).
- The percentage of voids filled with asphalt increases with increasing asphalt content (Chart 3).
- The curve for unit weight (density) of total mix generally rises with increased asphalt content and then falls, similar to the stability curve, except that the maximum unit weight normally occurs at a higher asphalt content than the maximum stability (Chart 4).

Table 3.07 Marshall Mix Design Criteria

Mix Criteria	Light Traffic Surface & Base		Medium Traffic Surface & Base		Heavy Traffic Surface & Base	
	Min.	Max.	Min.	Max.	Min.	Max.
Compaction, number of blows each end of specimen	35		50		75	
Stability, N (lbs.)	3336 (750)	—	5338 (1200)	—	8006 (1800)	—
Flow, 0.25 mm (0.01 in.)	8	18	8	16	8	14
Air Voids, percent	3	5	3	5	3	5
Voids in Mineral Aggregate (VMA), percent			*See Table 3.08*			
Voids Filled With Asphalt (VFA), percent	70	80	65	78	65	75

Notes:

1. All criteria, not just stability value alone, must be considered in designing an asphalt paving mix. Hot mix asphalt bases that do not meet these criteria when tested at 60°C (140°F) are satisfactory if they meet the criteria when tested at 38°C (100°F) and are placed 100 mm (4 in.) or more below the surface. This recommendation applies only to regions having a range of climatic conditions similar to those prevailing throughout most of the United States. A different lower test temperature may be considered in regions having more extreme climatic conditions.

2. Traffic classifications
   Light   Traffic conditions resulting in Design EAL ≤ 10^4
   Medium Traffic conditions resulting in a Design EAL between 10^4 and 10^6
   Heavy   Traffic conditions resulting in a Design EAL > 10^6

3. Laboratory compaction efforts should closely approach the maximum density obtained in the pavement under traffic.

4. The flow value refers to the point where the load begins to decrease.

5. The portion of asphalt cement lost by absorption into the aggregate particles must be allowed for when calculating percent air voids.

6. Percent voids in the mineral aggregate is to be calculated on the basis of the ASTM bulk specific gravity for the aggregate.

- Stability values generally increase as asphalt content increases, and then beyond a certain percentage of asphalt in the mixture stability decreases (Chart 5).
- Flow values increase with increasing asphalt contents (Chart 6).

►►*Determination of Design Asphalt Content*

The design asphalt content of the HMA is determined by considering the test data described above. First, determine the asphalt content where the percent air voids is equal to 4 percent or the designed target value. Then, evaluate all of the calculated and measured mix properties at this asphalt content by comparing them to the mix design criteria in Tables 3.07 and 3.08. If all of the criteria are met, then this is the design asphalt content. If all of the design criteria are not met, then some adjustment is necessary or the mix may need to be redesigned.

Table 3.08 Minimum VMA Requirements

Nominal Maximum Particle Size [1,2]		Minimum VMA, percent		
		Design Air Voids, percent [3]		
mm	openings/inches	3.0	4.0	5.0
1.18	No. 16	21.5	22.5	23.5
2.36	No. 8	19.0	20.0	21.0
4.75	No. 4	16.0	17.0	18.0
9.5	3/8	14.0	15.0	16.0
12.5	1/2	13.0	14.0	15.0
19.0	3/4	12.0	13.0	14.0
25.0	1.0	11.0	12.0	13.0
37.5	1.5	10.0	11.0	12.0
50	2.0	9.5	10.5	11.5
63	2.5	9.0	10.0	11.0

1. Standard Specification for Wire Cloth Sieves for Testing Purposes, ASTM E 11 (AASHTO M 92).
2. The nominal maximum particle size is one size larger than the first sieve to retain more than 10 percent.
3. Interpolate minimum voids in the mineral aggregate (VMA) for design air void values between those listed.

➤➤ Verifying Design Criteria

Using the data in Tables 3.07 and 3.08, we see from Chart 1 of Figure 3.07 that the asphalt content at four percent air voids is 4.7 percent. The values of the other mix properties are then checked to be certain that they satisfy the Marshall design criteria at that asphalt content. Referring again to the charts in Figure 3.07, we find that an asphalt content of 4.7 percent indicates these mix property values:

Stability (Chart 5) . . . . 10,400 N (2,300 lbs)
Flow (Chart 6) 9
Percent VFA (Chart 3) 70
Percent VMA (Chart 2) 13.9

These values are then compared with the values recommended by the Asphalt Institute for the Marshall design criteria of a heavily trafficked surface mix (Table 3.07). The stability value of 10,400 N (2,300 lbs.) exceeds the minimum criterion of 8,006 N (1,800 lbs). The flow value of 9 falls within the range of 8 to 14. The percent voids filled with asphalt (VFA) falls within the range of 65 to 75.

The minimum percentage of voids in the mineral aggregate (VMA) is checked using Table 3.08 and comparing it to the VMA of the specific aggregate gradation. Assume that the data in Figure 3.07 is for a gradation with a 3/4-in. nominal aggregate size. We then see that the VMA value of 13.9 exceeds the minimum of 13.0 required for a 3/4-in. mix containing four percent air voids.

►►Selecting a Mix Design

The mix design selected is usually the one that most economically meets all of the established criteria. However, the mix should not be designed to optimize one particular property. For example, mixes with abnormally high values of stability are often less desirable because pavements with such mixes tend to be less durable and may crack prematurely under heavy traffic. Any variations in design criteria should be allowed only under unusual conditions, unless the service behavior of a specific aggregate mixture indicates such a variant paving mix to be satisfactory.

Hveem Mix Design Method

The concepts of the Hveem Method of designing paving mixtures were developed by Francis N. Hveem, formerly Materials and Research Engineer with the California Division of Highways (Caltrans).

The Hveem method in its present form originated from investigations started by Caltrans in 1940. It involves determining an approximate asphalt content by Centrifuge Kerosene Equivalent test and then subjecting specimens at that asphalt content, and at higher and lower asphalt contents, to a stability test. A swell test on a specimen exposed to water is also made.

The Hveem Method is applicable to hot mix asphalt using penetration or viscosity graded asphalt cements and containing aggregate with maximum sizes of 25.0 mm (1 in.) or less. It is also applicable to mixtures using cut back asphalt. The method may be used for both laboratory design and field control of HMA.

The purpose of the Hveem Method is to determine the optimum asphalt content for a particular blend of aggregate. It also provides information about the properties of the resulting HMA. The Hveem Method utilizes a series of tests to determine optimum asphalt content:

- Centrifuge Kerosene Equivalent (CKE) test to determine an approximate asphalt content.
- Preparations of test specimens at the approximate asphalt content and at lower and higher asphalt contents.
- Stability test to evaluate resistance to deformation.
- Swell test to determine effect of water on volume change and permeability of specimen.

►►Conducting Hveem Mix Design

This is a general description of the procedures followed in the Hveem Mix Design Method. The complete and detailed procedure that must be followed is contained in ASTM D 1560 and D 1561/AASHTO T 246 and T 247. The Asphalt Institute publication *Mix Design Methods for Asphalt Concrete and Other Hot Mix Types*, Manual Series No. 2 (MS-2), details the procedures and equipment for conducting Hveem Mix Design.

As discussed in Chapter 2, different aggregates and asphalts have different characteristics. These characteristics have a direct impact on the nature of the pavement itself. The first step in the design method is to determine what qualities (stability, durability, workability, skid-resistance, etc.) the paving mixture must have, and, to select a type of aggregate and a compatible type of asphalt that will produce those qualities. Once this is done, test preparations can begin.

➤➤ Selection of Material Samples

The first step is to gather samples of the asphalt and the aggregate that will be used in the actual paving mixture. It is important that the asphalt samples have characteristics *identical* to those of the asphalt that will be used in the final HMA. The same is true of the aggregate samples. The reason is simple: data derived from the mix design procedures determine the job mix formula (JMF) for the actual paving mixture. The JMF can be accurate only if the ingredients tested in the laboratory have characteristics identical to those of the ingredients used in the final product. A wide variety of serious problems, ranging from poor workability of the mix to premature failure of the pavement itself, are historical results of variances between materials tested and materials actually used.

➤➤ Aggregate Preparation

Washed-Sieve Analysis The washed-sieve analysis involves these steps:

1. Each aggregate sample is dried to a constant weight.
2. Then each sample is washed thoroughly over a .075 mm (No. 200) sieve to remove any mineral dust coating the aggregate.
3. The washed samples are dried to a constant weight.
4. The dry weight of each sample is recorded. The amount of mineral dust is determined by subtracting the sample weight taken after washing from the sample weight before washing.

Determining Specific Gravity The specific gravity of any substance is the weight-volume ratio of a unit of that substance compared to the weight-volume ratio of an equal unit of water (see Materials section). The specific gravity of an aggregate sample is determined by comparing the weight of a given volume of the aggregate to the weight of an equal volume of water at the same temperature (Figure 3.03). The specific gravity is expressed in multiples of the specific gravity of water (which is always equal to 1). For example, an aggregate sample that weighs two-and-one-half times as much as an equal volume of water has a specific gravity of 2.5.

Calculating the specific gravity of the dry sample of aggregate (ASTM C 127 and C 128/ AASHTO T 84 and T 85) establishes a reference point for later specific gravity measurements taken to determine the proportions of aggregate, asphalt, and air voids in the test mixes.

Determining Surface Area of Aggregate The surface area of the aggregates is important in the Hveem Method because aggregate surface area, in addition to aggregate surface capacity, is used to estimate the asphalt content of the mixture. Surface area is determined by dry-sieving a carefully weighed aggregate sample and weighing the contents of each sieve. This information is converted into the estimated surface area of the sample by using the table of Surface Area Factors (Table 3.09). Surface area is expressed in terms of m^2/kg (ft^2/lbs) and varies inversely with the aggregate size.

Determining Surface Capacity of Aggregates An aggregate's surface capacity is its capacity for retaining a coating of asphalt. Table 3.10 demonstrates the calculation of surface area by this method.

The Centrifuge Kerosene Equivalent (CKE) procedure is used to determine an approximate asphalt content for an aggregate. The CKE procedure provides an index called a "K" factor that indicates relative particle roughness and surface capacity based on porosity.

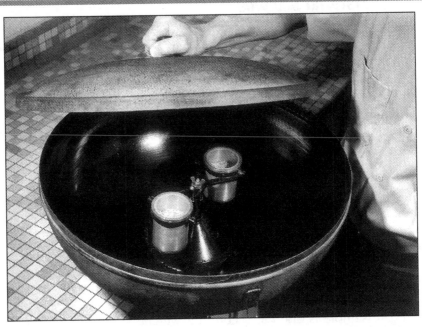

Figure 3.08 Saturating an Aggregate Sample with Kerosene (CKE Test)

Table 3.09 Surface Area Factors

Total Percent Passing Sieve No.	mm (No.)	4.75 (4)	2.36 (8)	1.18 (16)	0.60 (30)	0.30 (50)	0.15 (100)	0.075 (200)
Surface-Area Factor,* m²/kg (ft²/lb).	.41 (2)	.41 (2)	.82 (4)	1.64 (8)	2.87 (14)	6.14 (30)	12.29 (60)	32.77 (160)

* Surface area factors shown are applicable only when all the above-listed sieves are used in the sieve analysis.

Table 3.10 Example of Calculation of Surface Area

Sieve Size			Percent Passing	x	S.A. Factor		=	Surface Area	
19.0	mm	(3/4 in.)	100		.41	(2)		.41	(2.0)
9.5	mm	(3/8 in.)	90						
4.75	mm	(No. 4)	75		.41	(2)		.31	(1.5)
2.36	mm	(No. 8)	60		.82	(4)		.49	(2.4)
1.18	mm	(No. 16)	45		1.64	(8)		.74	(3.6)
0.60	mm	(No. 30)	35		2.87	(14)		1.00	(4.9)
0.30	mm	(No. 50)	25		6.14	(30)		1.54	(7.5)
0.15	mm	(No. 100)	18		12.29	(60)		2.21	(10.8)
0.075	mm	(No. 200)	10		32.77	(160)		3.28	(16.0)

Surface Area = 9.98 m²/kg (48.7 ft²/lb)

Figure 3.09 Rodding Mixture Sample in Mold

Figure 3.10 Mechanical Compactor Compacting Mixture Sample

The K factors are determined by tests that measure the amount of oil retained (Figure 3.08) on a coarse aggregate fraction [passing 9.5 mm (3/8 in.) sieve and retained on 4.75 mm (No. 4 sieve)], and the amount of kerosene retained on a fine aggregate fraction [passing the 4.75 mm (No. 4) sieve]. The factors are combined into a single factor representing the composite aggregate. The single factor along with the surface area of the aggregate is then used to determine an approximate asphalt content from a series of charts.

Preparing Mixture Test Samples Test samples of possible paving mixtures are prepared, each containing a slightly different amount of asphalt. The asphalt contents of the test samples are those suggested by data from the surface area and surface capacity tests. The aggregate proportion of the mixtures is formulated from the results of sieve analyses. Test specimens are prepared with the approximate asphalt content, 2 percent above it in 0.5 percent increments and one 0.5 percent below it. The sample preparation procedures are:

1. Asphalt and aggregate are heated and thoroughly mixed together until all aggregate particles are coated. This simulates the heating and mixing processes at a plant.
2. The resulting mixture is placed in an oven at 60°C (140°F) for fifteen hours to simulate storage of the mixture at the plant and the time lapse between production of the mixture and its placement. This allows the asphalt to age slightly and permits absorption of asphalt by the aggregate to occur, simulating the absorption that occurs during actual production and placement.
3. The mixture is heated to 110°C (230°F) to simulate compaction temperature.

4. The mixture is then placed in a compaction mold and tamped with a bullet-nosed rod (Figure 3.09). Rodding helps to ensure uniform compaction of the mixture under laboratory conditions.
5. A mechanical compactor is used to compact the mixture in simulation of compaction of the actual pavement (Figure 3.10).

▶▶ Hveem Test Procedure

The Hveem Method of Mix Design includes three test procedures: a stabilometer test, a bulk-density determination, and a swell test.

Stabilometer Test The stabilometer test is designed to measure the stability of the test mixture by subjecting it to specific stresses. The compacted test specimen is placed inside the stabilometer device, surrounded by a close-fitting rubber membrane (Figure 3.11). A vertical load is exerted on the sample and the resulting lateral (horizontal) pressure that develops is measured. The vertical pressure simulates the effects of pneumatic-tired loads repeated over a long period of time.

Stabilometer results depend largely on internal friction (resistance) of the aggregates and considerably less on the consistency of the asphalt. The Stabilometer test procedure is as follows:

1. The sample is heated to 60°C (140°F).
2. The sample is placed in the stabilometer (Figure 3.11).
3. The pressure in the stabilometer is raised to 34.5 kPa (5 psi).
4. Vertical loading is applied at a testing head speed of 1.3 mm/min. (0.05 in/min) until a 26.7 kN (6000 lbs.) load is reached.
5. Readings of the lateral pressure at specified loads are taken and recorded.
6. The vertical load is reduced to 4.45 kN (1000 lbs.) and a displacement measurement is made with the displacement pump.
7. The Hveem stability value for the sample is calculated using the information derived from the stabilometer test. The resulting stabilometer value (S) is based on the idea that HMA properties are somewhere between a liquid and a rigid solid. The value of S is derived from an arbitrary scale of 0 to 100, on which zero corresponds to a liquid having no measurable internal resistance to slowly applied loads and 100 corresponds to a hypothetical solid that transmits no measurable lateral pressure under a given load.

Air Voids Analysis The percent air voids are calculated from the bulk specific gravity of each compacted specimen and the maximum specific gravity of the paving mixture (zero voids). The latter is determined directly by a standard test (ASTM D 2041/AASHTO T 209) on an uncompacted sample of mix; or it can be calculated from the specific gravities of the asphalt and aggregate in the mix with an appropriate allowance made for the amount of asphalt absorbed by the aggregate. The bulk specific gravity of compacted specimens is determined by weighing specimens dry and saturated surface dry in air and immersed in water.

Swell Test Water is the enemy of all pavement structures. Consequently, a pavement mixture design must aim at giving the final pavement adequate water resistance to ensure its durability. The swell test (Figure 3.12) measures the amount of water that percolates into or through a specimen and the amount of swelling the water causes. It also measures the mixture permeability. The swell test procedure is as follows:

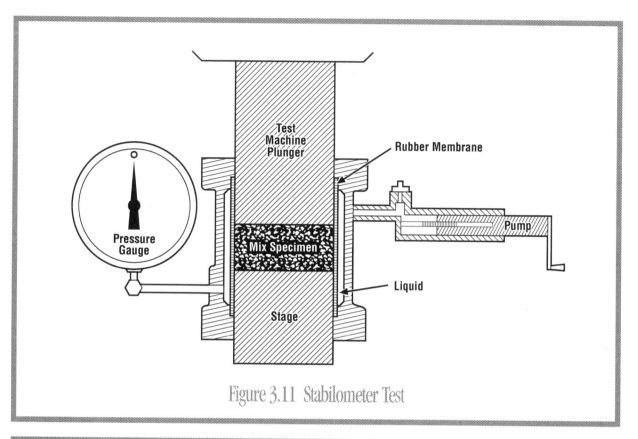

Figure 3.11 Stabilometer Test

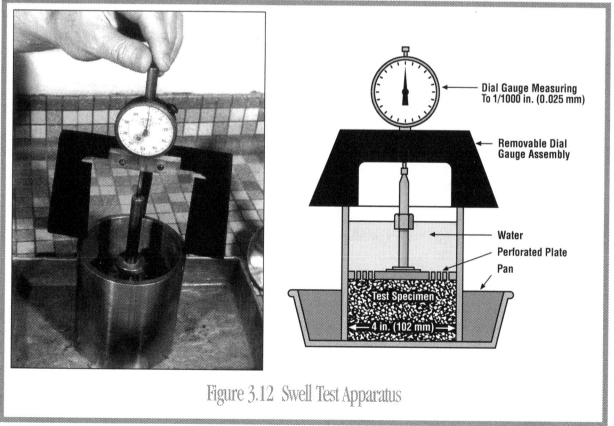

Figure 3.12 Swell Test Apparatus

1. The specimen, in its compaction mold, is placed in an aluminum pan and is covered with a perforated bronze plate.
2. A dial gauge is mounted over the test specimen so that its stem touches the bronze plate.
3. A specific amount of water is poured into the mold, directly onto the bronze plate.
4. The distance between the upper lip of the mold and the surface of the water is measured and recorded.
5. The specimen remains submerged for twenty-four hours.
6. A reading is taken from the dial gauge. This reading indicates how much the surface of the test sample has risen due to swelling.
7. The distance between the upper lip of the mold and the surface of the water is measured again. The difference between this measurement and that taken twenty-four hours earlier indicates how much water has seeped through the specimen. It is a measurement of the specimen's permeability.

➤➤Analyzing Hveem Test Results

The results of stabilometer testing, bulk density determination, and voids content measurements are recorded on a worksheet (Table 3.11) and are plotted on graphs as shown in Figure 3.13. These graphs are used to compare the characteristics of the test mixture samples.

The suitability of the mix designed by the Hveem method is determined on the basis of the asphalt content and aggregate gradation satisfying the criteria listed in Table 3.12.

The optimum asphalt content is normally the highest percentage the mixture will accommodate without reducing stability or void content below minimum values. The optimum asphalt content is determined by comparing three test mixture characteristics, namely stabilometer values, percentages of air voids, and tendency to flush or bleed. Pyramid charts (Figure 3.14) are used to make the comparisons and to determine the best test mixture.

The procedures used to determine the optimum asphalt content are as follows (also see Figure 3.15:

1. Record the asphalt contents used for preparing the mixture specimens in Step 1 of the pyramid. Record them in order of increasing amounts from left to right with the maximum asphalt content in the square on the right.
2. Select from Step 1 the three highest asphalt contents that do not exhibit moderate or heavy surface flushing and record them in Step 2. Surface flushing and/or bleeding is considered "slight" if the surface has a slight sheen. It is considered "moderate" if sufficient free asphalt is apparent to cause paper to stick to the surface but no distortion is noted. Surface flushing is considered "heavy" if there is sufficient free asphalt to cause surface bubbling or specimen distortion after compaction.
3. Select from Step 2 the two highest asphalt contents that provide the specified minimum stabilometer value and enter them in Step 3.
4. Select from Step 3 the highest asphalt content that has at least 4.0 percent air voids and enter it in Step 4. This is the optimum asphalt content. However, if the maximum asphalt content used in Step 1 is the asphalt content entered on Step 4, additional specimens with increased asphalt content must be prepared and a new optimum asphalt content determination made. This is because a greater asphalt content than the maximum tested might prove to be better for the pavement design.

Sp. Gr. Asp. Cem. 1.012 Asp. Cem. AC-10
Avg. Bulk Sp. Gr. Agg. = 2.760

Lab. No. for ASP Cem. Used: 53-0741
Lab. Nos. for Agg. Used: 53-1252; 53-1253

Gradation, CKE, and Percent Asphalt

Sieve Size mm (in. or No.)	19.0 (3/4)	12.5 (1/2)	9.5 (3/8)	4.75 (4)	2.36 (8)	1.18 (16)	0.60 (30)	0.30 (50)	0.15 (100)	0.075 (200)
Specification Limits	100	100 80	90 70	70 50	50 35		29 18		16 8	10 4
Percent Passing	100	91	76	60	42	32	23	16	12	6
S.A. Factors			.41	.41	.82	1.64	2.87	6.14	12.29	32.77
Surface Area, m²/kg*			.41	0.25	0.34	0.53	0.66	0.98	1.48	1.97

CKE: FA = 2.8; CA = 2.8 Kf = 1.0; Kc = 1.3; Km = 1.0; Total Surface Area = 6.62 m²/kg (32.3 ft²/lb)
Estimated Percent Asphalt Cement by Weight of Aggregate using CKE Tests only = 5.5
Recommended Percent Asphalt Cement by Weight of Aggregate using Mix Design Criteria = 5.0

Specimen Identification	A	B	C	D
Percent Asphalt Cement By Weight of Aggregate	5.0	5.5	6.0	6.5
Percent Asphalt Cement By Weight of Mix	4.76	5.21	5.66	6.10
Weight in Air, grams	1211.0	1223.3	1230.8	1235.9
Weight in Water, grams	714.9	723.8	727.6	733.3
Bulk Volume, cc.	496.1	499.5	503.2	502.6
Bulk Specific Gravity	2.441	2.449	2.446	2.459
Maximum Specific Gravity	2.559	2.540	2.522	2.504
Air Voids, Percent	4.6	3.6	3.0	1.8
Unit Weight (kg/m³), pcf.	2.439 (152.3)	2.448 (152.8)	2.446 (152.6)	2.457 (153.4)

Total Load kN	lbs.	Unit Load MPa	psi.	Stabilometer			
2.22	500	0.28	40	9	9	9	10
4.45	1000	0.55	80	12	12	15	16
8.90	2000	1.10	160	15	16	24	26
13.34	3000	1.65	240	21	22	30	38
17.79	4000	2.21	320	28	30	42	55
22.24	5000	2.76	400	36	39	55	83
26.69	6000	3.31	480	50	52	62	105

	A	B	C	D
Displacement, turns	2.40	2.50	2.46	2.50
Stability Value	48	45	36	25

Jones
Inspector

* Surface Area: 1 m²/kg = 4.8824 ft²/lb

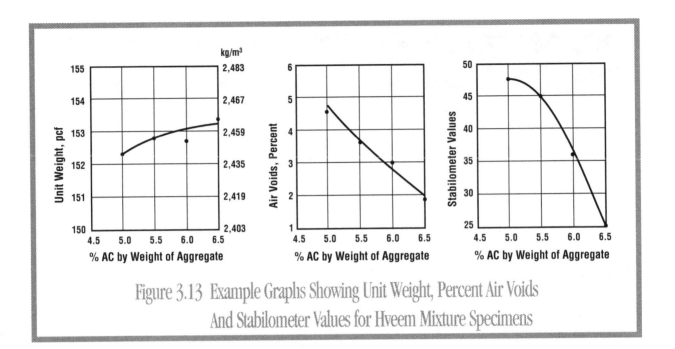

Figure 3.13 Example Graphs Showing Unit Weight, Percent Air Voids And Stabilometer Values for Hveem Mixture Specimens

Table 3.12 Hveem Mix Design Criteria

Traffic Category	Heavy		Medium		Light	
Test Property	Minimum	Maximum	Minimum	Maximum	Minimum	Maximum
Stabilometer Value	37	—	35	—	30	—
Swell	less than 0.762 mm (0.030 in.)					

Notes:
1. Although not a routine part of this design method, an effort is made to provide a minimum percent air voids of approximately 4 percent.
2. All criteria, and not stability value alone, must be considered in designing an asphalt paving mix.
3. Hot mix asphalt bases, which do not meet the above criteria when tested at 60°C (140°F) should be satisfactory if they meet the criteria when tested at 38°C (100°F) and are placed 102 mm (4 inches) or more below the surface. This recommendation applies only to regions having climatic conditions similar to those prevailing throughout most of the United States. Guidelines for applying the lower test temperature in regions having more extreme climatic conditions are being studied.

Superpave Mix Design

The Superpave (short for Superior Performing Asphalt Pavements) asphalt mix design system is a product of the asphalt research conducted by the Strategic Highway Research Program (SHRP). Superpave is a performance-based system for specifying asphalt binders and mineral aggregates, performing asphalt mixture design, and analyzing pavement performance. The term "asphalt binder" is used in order to include both modified and unmodified asphalt cements.

Superpave includes an asphalt binder specification that uses new binder physical property tests; a series of aggregate tests and specifications; a hot mix asphalt (HMA) design and analysis system using the Superpave gyratory compactor; and computer software to integrate the system components. As with any design process, field control measurements are still necessary to ensure the field produced mixtures match the laboratory design. These procedures are covered in Chapter 7.

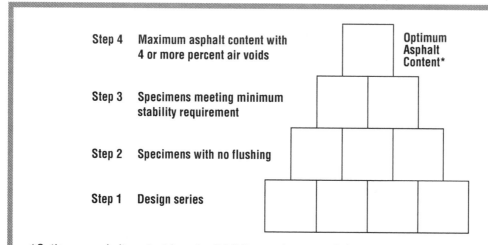

Step 4   Maximum asphalt content with
          4 or more percent air voids

Step 3   Specimens meeting minimum
          stability requirement

Step 2   Specimens with no flushing

Step 1   Design series

Optimum
Asphalt
Content*

*Optimum asphalt content is not valid if the maximum asphalt content used in the design series (Step 1) is the asphalt content arrived at in Step 4. In this event, additional specimens must be prepared with increased asphalt content in 0.5 percent increments and a new analysis made.

Figure 3.14 Pyramid Chart Used to Determine Optimum Asphalt Content

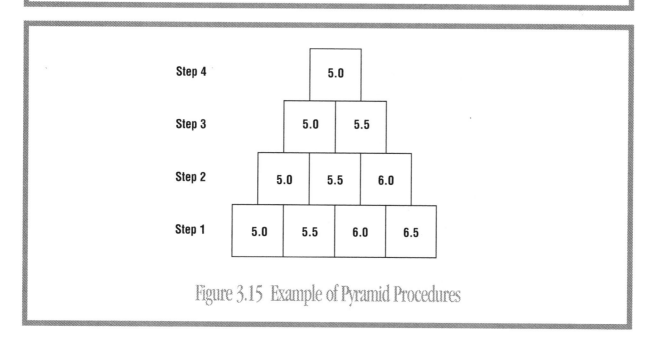

Figure 3.15 Example of Pyramid Procedures

The Superpave system was developed to provide the tools necessary to design asphalt mixes that will perform better under heavy traffic and extreme temperatures. The goal is to provide pavements that are resistant to rutting, fatigue cracking, and low temperature cracking. The Superpave performance-graded (PG) binder specifications take into account the range of temperatures and the aging which the asphalt binder will endure during its performance life. A unique feature of the specification is that the tests are actually performed on the asphalt at the temperatures the pavement must withstand and after simulated aging.

This discussion will briefly detail the Superpave mix design system. Complete details for conducting Superpave mix design can be found in the Asphalt Institute publication *Superpave Mix Design*, Superpave Series No. 2 (SP-2).

There are four major steps in the current testing and analysis process:

1. Selection of materials (asphalt binder and aggregates).
2. Selection of a design aggregate structure.
3. Selection of a design asphalt binder content.
4. Evaluation of moisture sensitivity of the design mixture.

➤➤ Selection of Materials

Selection of materials consists of determining the traffic and environmental factors for the paving project. From that, the performance grade of asphalt binder required for the project is selected. Aggregate requirements are determined based on traffic level and the depth of the HMA in the pavement. Materials are selected based on their ability to meet or exceed the established criteria.

Asphalt Binder Performance-graded (PG) asphalt binders are selected based on the climate in which the pavement will serve. The distinction among the various binder grades is the specified maximum and minimum pavement temperatures for the design criteria, in degrees Celsius (°C).

PG binder grades contain a high temperature grade and a low temperature grade. For example, a PG 58-22 is designed to withstand the conditions of an environment where the average seven day maximum pavement temperature does not exceed 58°C and the minimum pavement temperature is not less than -22°C. PG asphalt binders are classified into standard grades in six-degree increments as shown in AASHTO MP1 (Appendix E).

The PG binder grades requirements can be determined one of three ways: using actual pavement temperature measurements; using air temperature measurements that are converted to pavement temperatures; or by geographic areas developed within a particular agency. Many agencies have used the last two methods to determine the "standard" PG grades for use within their jurisdiction.

The high temperature PG grade can also be adjusted for heavier traffic conditions and speed of traffic. Depending on the amount and speed of truck traffic, the high temperature grade can be increased one or two levels.

Aggregates Superpave aggregate requirements include "consensus properties" that vary depending on the amount of traffic in the design project, as well as the depth within the pavement the HMA is to be used. The consensus properties are:

- Coarse aggregate angularity
- Fine aggregate angularity
- Flat and elongated particles
- Clay content

It is recommended that agencies also use "source properties" in addition to consensus properties. These properties are considered to be source specific, that is they are specified by each agency in relation to the aggregates available in the area. The source properties are:

- Toughness
- Soundness
- Deleterious materials
- Others as necessary

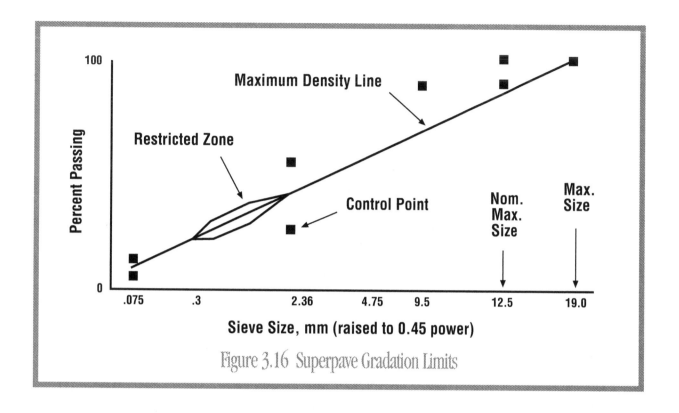

Figure 3.16 Superpave Gradation Limits

▶▶ Selection of Design Aggregate Structure

The selection of the Superpave design aggregate structure is a trial-and-error process similar to other design procedures. In the case of Superpave, the selection of the design aggregate structure includes not only determining the aggregate gradation, but also the volumetric properties of the mixture after it has been compacted in the Superpave gyratory compactor.

Superpave uses the 0.45 power gradation chart to define a permissible gradation (Figure 3.16). An important feature of this chart is the maximum density gradation. This gradation plots as a straight line from the maximum aggregate size through the origin. As discussed in Chapter 2, the maximum density gradation represents a gradation in which the aggregate particles fit together in the densest possible arrangement. This is a gradation to avoid because there is very little aggregate space to develop sufficiently thick asphalt films for a durable mixture.

To specify Superpave aggregate gradation, two additional features are added to the 0.45 power chart: control points and a restricted zone. Control points put limits on the aggregate gradation. They are placed on the nominal maximum size, an intermediate size (2.36 mm), and the dust size (.075 mm). A maximum control point is also placed on the size below the nominal maximum size.

The restricted zone resides along the maximum density gradation between the intermediate size (either 4.75 or 2.36 mm, depending on the maximum size) and the 0.3 mm size. It forms a band through which it is recommended that gradations should not pass. Gradations that pass through the restricted zone, starting from below the zone, have often been called "humped gradations" because of the characteristic hump in the grading curve that passes through the restricted zone.

A humped gradation often results in a tender mix, which is a mixture that is difficult to compact and has reduced resistance to rutting during its performance life. Gradations that violate the restricted zone possess weak aggregate skeletons that depend too much on asphalt binder stiff-

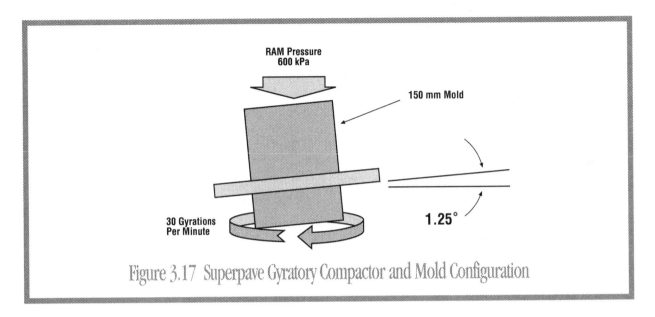

Figure 3.17 Superpave Gyratory Compactor and Mold Configuration

ness to achieve mixture strength. These mixtures are also very sensitive to asphalt content and can easily become plastic. Superpave recommends that mixtures be graded below the restricted zone, especially for high traffic projects.

For Superpave mix design, several trial blends are selected that meet the gradation criteria, generally three gradations ranging from coarse to fine. These blends are mixed with a trial asphalt content and then compacted in the Superpave gyratory compactor (SGC) (Figure 3.17). The SGC compacts the mixture in a mold through a combination of constant vertical pressure and a constant angle of gyration applied around the mold 30 times per minute. The angle of gyration and the vertical pressure create a kneading action that compacts the asphalt mixture specimen. The appropriate number of gyrations increases as traffic increases.

As the specimen compacts, its height is recorded so that the specimen's rate of densification can be determined. After compaction is completed, the specimen is removed from the mold, and the bulk specific gravity is determined. The maximum theoretical specific gravity of the mixture is also determined. These two measurements and the change in height during compaction are used to generate a densification curve (Figure 3.18) comparing each mixture's percentage of maximum theoretical specific gravity (the inverse of air voids) versus the number of gyrations. These curves help determine whether the mix may show tender mix characteristics during construction, as well as rutting characteristics later on during the mixture's life.

In addition to compaction characteristics, the HMA trial blends are also evaluated for mixture volumetrics. The volumetric properties evaluated are:

- Air voids
- Voids in the mineral aggregate (VMA)
- Voids filled with asphalt (VFA)

All mixtures are evaluated at an air void content of four percent. The VMA criterion changes based on the nominal maximum size of the HMA, with finer mixtures requiring more VMA. The VFA criterion changes as a function of traffic. Increasing traffic results in a lower required range of VFA. Lower volume roads require higher VFA ranges to increase the roadway's durability.

The dust proportion (ratio of the percent passing the 0.075 mm [No. 200] sieve to the effective asphalt content_ is also evaluated in order to obtain the most desirable mixture. The Superpave requirement for dust proportion is within the range of 0.6 to 1.2.

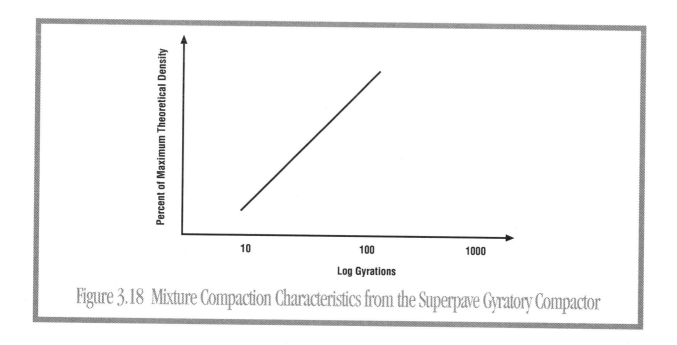

Figure 3.18 Mixture Compaction Characteristics from the Superpave Gyratory Compactor

The trial aggregate blend that best meets all compacted mixture criteria should be selected as the design aggregate structure. The blend is then further evaluated by varying the asphalt content.

►►Selection of Design Asphalt Content

After an aggregate blend and an estimated design asphalt content meeting the Superpave requirements have been developed, the final design asphalt content is determined. This step consists of varying the amount of asphalt binder in test specimens, using the design aggregate structure, in order to obtain volumetric and compaction properties that comply with the mixture criteria. The process also verifies the asphalt content results estimated from the previous step. In addition, the sensitivity of volumetric and compaction properties of the design aggregate structure as compared to asphalt content is evaluated here.

The procedure involves mixing and compacting design aggregate structure specimens at several asphalt contents. Generally four asphalt binder contents are used, centered about the estimated design asphalt content. Values of air voids, VMA, and VFA are determined for each asphalt content. Graphs are generated showing the variation in these mixture property values as the asphalt content varies (Figure 3.19). The design asphalt binder content is selected as the value that corresponds with four percent air voids. The values of VMA and VFA are then checked at that asphalt content to ascertain that the appropriate Superpave criteria are met. Other Superpave criteria are also checked before making the final design asphalt binder content selection.

►►Evaluation of Moisture Sensitivity

The final step in Superpave mix design is an evaluation of the moisture sensitivity of the design HMA. The loss of bond, or stripping, caused by the presence of moisture between the asphalt and aggregate is a problem in some areas and can be severe in some cases. Many fac-

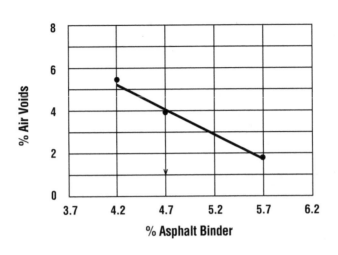

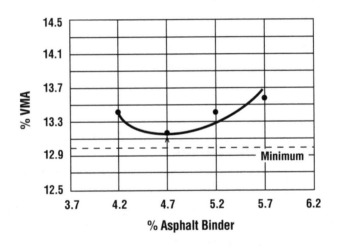

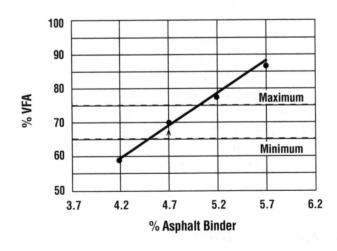

Figure 3.19 Air Voids, VFA and VMA vs. Asphalt Content

tors such as aggregate characteristics, asphalt characteristics, environment, traffic, construction practices and drainage can contribute to stripping.

Currently, AASHTO T 283 is used to evaluate HMA for stripping. Research is being conducted to develop an improved test. In the test, two subsets of test specimens are produced at an air void content of seven percent, plus or minus one percent. One subset is moisture conditioned by vacuum saturation to a constant degree of saturation in the range of 55 to 80 percent. This is followed by an optional freeze cycle. The final conditioning step is a hot water soak. No conditioning is conducted on the other subset.

After conditioning, both subsets are tested for indirect tensile strength. The test result reported is the ratio of tensile strength of the conditioned subset to that of the unconditioned subset. This ratio is called the "tensile strength ratio," or TSR. Mixtures with less than an 80 percent TSR are considered moisture susceptible.

Purpose and Objective

An asphalt plant is an assembly of mechanical and electronic equipment where aggregates are blended, dried, heated, and mixed with asphalt to produce hot mix asphalt (HMA) meeting specified requirements. Asphalt plants vary in mixing capacity, and may be stationary (located at a permanent location) or portable (moved from job to job). In general, every plant can be categorized as either a batch plant (Figure 4.01) or a drum mix plant (Figure 4.02).

Regardless of the type of hot mix plant, the basic purpose is the same – to produce HMA containing the specified proportions of asphalt and aggregate. Both batch plants and drum mix plants are designed to accomplish this purpose.

The objective of this chapter is to provide understanding and knowledge of:

- The function of an asphalt plant.
- The two basic types of asphalt plants and the major components of each.
- The proper procedures for handling, storing, and sampling HMA components (aggregate, asphalt, and mineral filler).
- The operation of cold aggregate feed systems.
- Basic sampling and testing procedures for checking HMA characteristics
- Items to be checked in a visual inspection of HMA.
- The items that should appear in plant records.
- Safety considerations necessary for safe and efficient plant operation.

General Information

The basic operations involved in producing HMA are the same regardless of the plant type. These operations include:

1. Proper storage and handling of HMA component materials at the mixing facility.
2. Accurate proportioning and feeding of the cold aggregate to the dryer.
3. Effective drying and heating of the aggregate to the proper temperature.
4. Efficient control and collection of the dust from the dryer.
5. Proper proportioning, feeding and mixing of asphalt with heated aggregate.
6. Correct storage, dispensing, weighing, and handling of finished HMA.

There are many styles and types of equipment that can effectively accomplish these steps. It would be impossible to cover all of these possibilities in this publication. Therefore, the focus will be on the generally accepted practices for producing quality HMA. This chapter will present plant operation procedures following these basic steps.

The procedures listed above are common to all types of mixing facilities. The basic difference between batch and drum mix plants is in how they mix asphalt and aggregate after it has been dried and heated (Step 5). Batch plants screen and separate the aggregate into separate bins after it has been

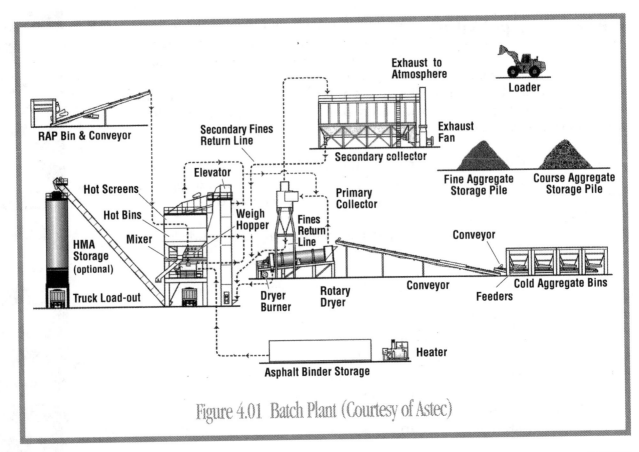

Figure 4.01 Batch Plant (Courtesy of Astec)

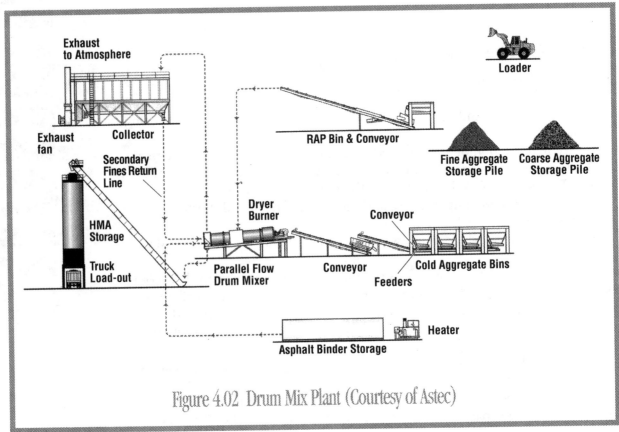

Figure 4.02 Drum Mix Plant (Courtesy of Astec)

dried, and then recombine the aggregate and mix it with the asphalt, one batch at a time in a separate mixer. Drum mix plants dry the aggregate and blend it with asphalt in a continuous process within the dryer drum; hence the name drum mixer.

Common to all HMA facilities is the importance of uniformity and balance of plant operations and materials and. Uniformity ensures that the HMA is consistently produced to meet job specifications. It encompasses uniformity of materials, uniformity of material proportioning, and continuous, uniform operation of all plant components. Changes in material characteristics or proportions and intermittent stops and starts in plant operations make it extremely difficult to produce quality HMA.

Balance involves careful coordination of all elements of production and placement. Balancing material quantities to plant production and balancing plant production and pavement placing operations guarantee a continuous, uniform production and placement effort.

Uniformity and balance are accomplished by careful preparation. Materials must be readily available, sampled, tested and approved, and plant components must be carefully inspected and calibrated before production begins.

Material Handling

The quality of the HMA produced can only be as good as the materials going into the plant. One of the basic necessities of ensuring quality HMA production is that an adequate supply of suitable materials be available prior to and during mixing operations. These sections discuss the handling and control of both asphalt and aggregate. The principles presented are common to all HMA plants.

It is important to understand that in modern mixing plants, aggregate gradation and quality are assured at the quarry, not at the mixing facility. Batch plants are only able to make minor adjustments in gradation, and drum mix plants can make none at all. The plant has no equipment that detects or corrects for variations in aggregate quality or gradation.

➤➤Aggregate Storage and Handling

Aggregates must be handled and stored in a manner that avoids contamination and minimizes degradation and segregation, and avoids contamination. The stockpile area should be clean and stable to prevent aggregate contamination. Materials should be stockpiled on a free-draining grade to prevent accumulation of moisture. Storing aggregate stockpiles under roof is an increasingly used option to keep precipitation off of the aggregates.

Steps must be taken to prevent intermingling of different aggregates. Stockpile areas must include enough space for clear separation of aggregates, or use bulkheads between stockpiles. If bulkheads are used, they should extend to the full height of the stockpile to prevent overflowing and subsequent contamination.

The primary concern with the handling and stockpiling of aggregates is segregation, and the methods used to control it depend on the nature of the material. Coarse aggregates composed of several particle sizes require more careful handling than fine-graded aggregates (such as sands and fine screenings) and one-sized coarse aggregates. Sands, crushed fine aggregate and single-size aggregates – especially the smaller sizes – can be handled and stockpiled in almost any manner with little, if any, segregation.

Aggregate blends, however, require special handling. For example, if material containing both coarse and fine particles is placed in a stockpile with sloping sides (a cone shape), segre-

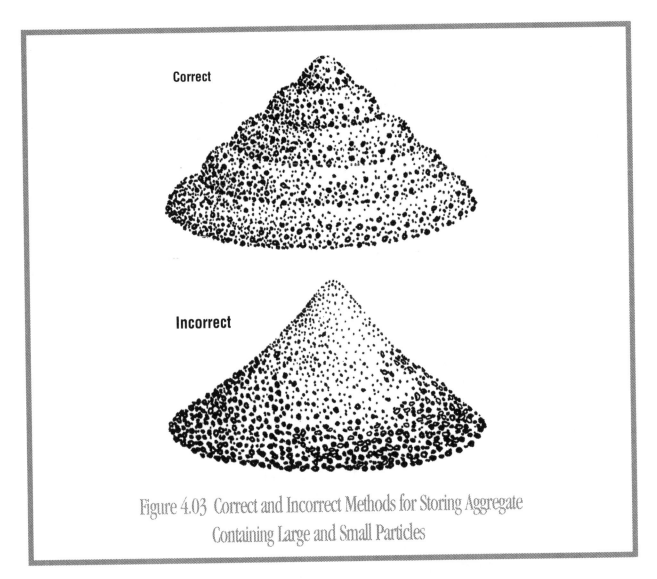

Correct

Incorrect

Figure 4.03 Correct and Incorrect Methods for Storing Aggregate
Containing Large and Small Particles

gation is sure to occur as the larger particles roll down the slope as shown in Figure 4.03 (incorrect). Building a stockpile in layers as shown in Figure 4.03 (correct) can minimize such segregation.

Some choices are available when stockpiling aggregates. *Horizontal stockpiling* (Figure 4.04) with end-dump aggregate trailers is probably the most commonly used method for stockpiling aggregates at a hot mix facility. Each end-dump trailer load should be dumped into an individual pile. Pushing or dumping aggregate over the side of the stockpile should be avoided as it may result in segregation. Steel-tracked and some rubber-tired equipment should be kept off the stockpiles as much as possible to prevent aggregate breakage, fines generation and degradation of the stockpile, especially if the aggregate is soft (Figure 4.05).

The use of radial stacking conveyors allows more material to be stockpiled over a smaller area by raising the elevation of the stockpile. Radial stackers can be used at truck dumping stations or at rail car off-loading areas.

Proper use of a radial stacker includes raising the conveyor slowly after moving it horizontally to cause the stockpile to grow vertically. Segregation can occur if a stacker is allowed to drop aggregate from a great height.

Figure 4.04 Building a Horizontal Stockpile

Figure 4.05 Equipment Working on Top of a Stockpile

Table 4.01 Size of Samples Adequate for Routine Grading and Quality Analysis per AASHTO T 2

Nominal Maximum Particle Size		Approximate Minimum Weight of Field Samples*	
Fine Aggregate			
mm	U.S. Customary	kg	lb.
2.36	No. 8	10	25
4.75	No. 4	10	25
Coarse Aggregate			
12.5	1/2 in.	15	35
9.5	3/8 in.	10	25
19	3/4 in.	25	55
25	1 in.	50	110
37.5	1-1/2 in.	75	165
50	2 in.	100	220
63	2-1/2 in.	125	275
75	3 in.	150	330
90	3-1/2 in.	175	385

* The samples prepared for tests shall be obtained from the field sample by quartering or other suitable means to insure a representative portion.

▶▶Aggregate Sampling

Stockpiled aggregates should be sampled and tested at regular intervals to ensure that a uniform gradation is maintained. Each stockpile sample must be a composite or mixture of aggregates taken from different levels of the pile (top, middle, and near the bottom). A wood or metal shield shoved vertically into the pile just above the point of sampling will prevent loose aggregate particles from sliding down into sampling areas.

A square-shaped shovel with turned-up edges that form a scoop should be used for sampling. The shovel blade is inserted horizontally into the stockpile so as to remove a scoop-full of material. A straight edge can be used to strike off the shovel to reduce segregation from spillage, and maintain a representative sample. The aggregate sample is placed in a clean bucket or other sturdy container. It is important to obtain the sample from various locations of the stockpile. It is also important that sampling locations not be in a vertical line. They should be staggered around or within the pile to ensure a representative sample.

Aggregate samples can also be taken from a conveyor belt or a sampling chute. These samples, if taken properly, are usually more representative than stockpile samples. The conveyer belt must be stopped and the sample must include material covering the full width of the belt in order to obtain a representative sample. The entire flow of aggregate from the belt must enter the sampling chute.

Table 4.01 lists aggregate sample sizes, by weight, based on nominal maximum particle size.

Mechanical Splitting Method

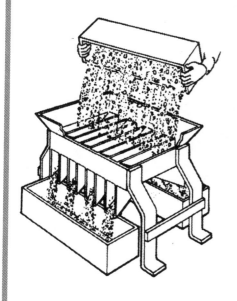

Quartering Method

Sample Quartering – Sample sizes can also be reduced by quartering. The quartering method should be used when splitters are not available. Quartering simply requires a quartering cloth and a stick or rod – and is done as follows:

1. Pour contents from sample bucket onto quartering cloth.

2. Level sample on quartering cloth – using rod.

3. Insert rod under the middle of the quartering cloth and lift both ends of rod to divide the sample into two equal parts.

4. Repeat step 3, dividing the sample into four parts.

5. Retain any two diagonally opposite parts for testing.

If the sample is still not small enough, repeat the quartering procedure using either of the two diagonally opposing halves.

Figure 4.06 Aggregate Sample Reduction

After obtaining an aggregate sample, it is usually necessary to reduce the sample size for testing convenience. Because reducing a sample presents an opportunity for segregation, care must be taken to preserve the samples integrity. Two sample reduction methods are illustrated in Figure 4.06. A mechanical splitter is usually preferred for coarse aggregate or dry, fine aggregate. Quartering is the best method when the aggregate sample is wet. ASTM C 702 and AASHTO T 248 describe both methods in detail.

Random (or statistical) sampling is not included in this chapter. The procedure for this type of sampling is included in Chapter 7 in the section titled Sampling and Testing Plan for Quality Control. Additional information is located in Appendix C. ASTM D 3665, Standard Practices for Random Sampling, provides information on this subject.

➤➤ Storage and Handling of Asphalt

Enough asphalt must be stored at the plant to maintain uniform plant operation, especially when taking delayed shipments and testing time into account. Most plants have at least two asphalt tanks – one working tank and one or more storage tanks. Some plants are using vertical storage tanks to better handle modified asphalt (Figure 4.07). When more than one grade of asphalt is required for production, at least one tank will be needed for each grade.

Figure 4.07 Asphalt Storage Tanks (Courtesy of Heatec)

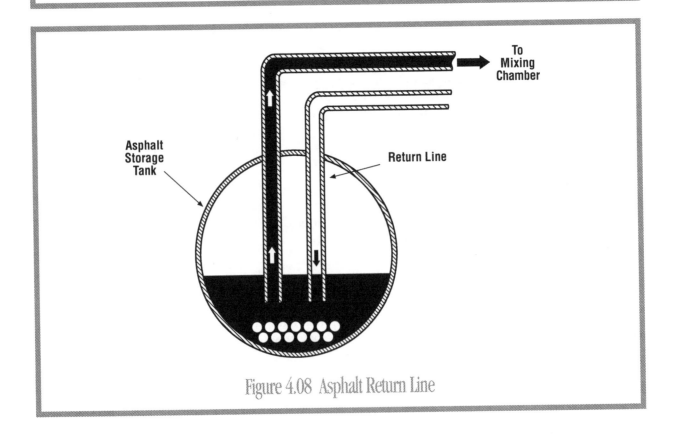

Figure 4.08 Asphalt Return Line

Figure 4.09 Sampling Valve in Delivery Line to Mixing Chamber

Asphalt storage tanks should be calibrated so that the amount of material remaining in the tank can be determined at any time. They must be heated to keep the asphalt fluid enough to move through the delivery and return lines. Heating is done electrically or by circulating hot oil through coils in the tank. Regardless of the heating method used, an open flame must never come in contact with the tank or its contents.

Where circulating hot oil is used, the oil level in the heating unit reservoir should be checked frequently. A drop in the level could indicate hot oil leakage into the tank resulting in contamination of the asphalt.

All transfer lines, pumps and weigh buckets must also have heating coils or jackets to keep the asphalt hot. One or more thermometers should be placed in the asphalt feed system to monitor asphalt temperature.

Return lines discharging into the storage tank should be submerged below the asphalt level at all times to prevent the asphalt from oxidizing during circulation (Figure 4.08). In order to break the vacuum in the lines when the pump is reversed, and to clear the lines, two or three vertical slots can be cut in the return line within the tank, but above the high level mark of the stored asphalt.

Any material (especially modified asphalt) remaining in a storage tank should be purged before adding a different grade. Different formulations of even the same grade may not be compatible and should never be combined in the same tank. The asphalt supplier should be contacted when considering the intermixing of asphalt.

Figure 4.10 Asphalt Sampling Device for Storage Tank (Courtesy of Heatec)

▶▶ *Sampling Asphalt*

Normally, asphalt samples are taken from a sampling valve in the delivery line to the mixing chamber (Figure 4.09). Occasionally, samples will be taken from a sampling device on a vehicle tank or storage tank (Figure 4.10). A few important rules to follow when sampling asphalt are:

- All safety precautions for handling and storing hot asphalt should be followed.
- Asphalt binder is very hot when sampled, so protective clothing (gloves, face shield, long-sleeved shirt) should be worn to protect from burns.
- To ensure that samples are representative of the entire shipment, they should be taken from the sampling valves provided for that purpose.
- Sampling methods are described in ASTM D 140 and AASHTO T 40.
- Only new, clean, and dry sample containers are to be used.
- At least a quart of asphalt binder should be allowed to drain out of the valve before taking samples. This cleans out the valve and the lines and helps provide a representative sample.
- Filled containers should be sealed immediately with clean, dry, tight fitting lids. Any spilled material should be wiped from the container with a clean, dry cloth – NEVER with a cloth dipped or soaked in solvent.
- All containers must be labeled clearly. Container lids should also be labeled because once a lid is removed, it will be necessary to match it to the appropriate container. Containers should be labeled with a permanent marking pen. Tags should be used only when there is no danger of their being lost in transit.

Figure 4.11 Silo for Feeding Mineral Filler

➤➤ *Mineral Filler*

Storage and Handling Mineral filler is subject to caking or hardening when exposed to moisture. Therefore, separate storage should be provided to protect from dampness. In plant operations where mineral filler usage is high, a bulk storage silo is often used (Figure 4.11). Such a system can have either a pneumatic or mechanical device for feeding filler into the plant. In pneumatic systems, mineral filler is entrained in an airstream and is handled as a fluid, offering accurate control and eliminating plugging.

The pneumatic system generally consists of a receiving hopper, screw conveyor, dust-tight elevator, and silo. The hopper and elevator load the silo from which filler is metered into the plant. The silo may also be loaded directly from suitably equipped filler transport trucks. The mineral filler is normally introduced into the mix in the weigh hopper of the batch plant. In some plants a separate weighing system may be provided. In a drum plant the mineral filler is introduced pneumatically through a pipe located where the asphalt is introduced.

A bag-feeding, mineral filler system may also be used. This system consists of a ground-mounted feeder, dust-tight elevator, surge hopper, vane feeder or screw conveyor, and an overflow chute.

In both bulk and bag systems, final metering of the filler into the mix is accomplished through a variable-speed vane or screw feeder, or belt feeder depending on the material to be handled and the capacity required. In each case, the filler feed mechanism is interlocked with the aggregate and asphalt feed mechanisms to ensure uniform proportioning.

Filler handling also involves a plant's dust collection system. Dust collectors are designed to capture fines escaping from the aggregate mixture and return them to the plant for incorporation into the mix.

Where excess fines are encountered in the raw aggregate feed, a bypass system can be used to receive the filler collected by the dust collector. The required amount of filler is then fed back to the mix and any surplus amount is diverted to a storage bin.

Hydrated lime or portland cement is sometimes used as mineral filler. Hydrated lime is also used as an antistripping agent. A pneumatic handling system is generally used when either of these materials is incorporated dry into the mix. When used as an antistripping agent, hydrated lime can be added to the moist aggregate or made into a slurry and then thoroughly mixed in an approved pugmill. Generally the additive-aggregate mixture is fed directly into the plant after mixing. However, a lime slurry mixture may be stockpiled before introduction into the plant for mixing.

Control of Feeding When the mineral filler is added to the mix, its proportioning must be exact. Consequently, the flow of filler into the plant must be carefully controlled and frequently checked.

The percentage of filler entering the hot mix can be calculated by simply measuring the amount of filler consumed by the plant during production of a given amount of HMA.

Modern hot mix facilities have load cells mounted on storage silo supports, allowing the monitoring of the amount of material in storage at any given time. If this method is not available, close and frequent checks on the calibration of the feeding and weighing mechanisms can be taken.

➤➤ Material Records

Aggregates Certain information should be recorded as aggregates are received at the plant site. This includes a description of the material, the date and quantity delivered, and whether the material has been tested and approved before delivery. If it has been approved, the test identification number should be recorded and samples taken as necessary to verify the test data. The size and frequency of such check samples will vary with the policy of the specifying agency.

If the material has not been tested before delivery, random samples should be obtained and tested to ensure compliance with all specifications. At a minimum, tests should be made for gradation (sieve analysis) and cleanliness (washed sieve analysis) for coarse and fine aggregates, and sand equivalency for fine aggregate. Samples are also frequently taken for absorption, specific gravity, abrasion, and tendency to strip (affinity for asphalt).

Records for untested materials should include:

- Name of material producer.
- Location of supply source.
- Quantity delivered.
- Quantity represented by each sample.

Asphalt In most instances, asphalt comes from a pretested source and is accepted by certification. Even so, a record must be kept of all deliveries of asphalt binder to the HMA plant. The records should include the following information:

- Asphalt binder grade
- Producers name and location or shipping point
- HMA Plant and project identification
- Date of delivery
- Delivery invoice number
- Certification number
- Asphalt quantity by weight or calculated volume based on direct measurement

Where the volume is calculated, the following information must be furnished:

a) Calibration chart identification
b) Beginning measurement (before unloading)
c) Ending measurement (after unloading)
d) Temperature of asphalt when measured
e) Temperature correction factor for converting to equivalent liters at 15°C (gallons at 60°F)
f) Equivalent liters (gallons)
g) Proper identification of check samples taken

Similar records should be kept on all other materials (such as mineral filler additives) that are to be incorporated into the mix.

►►Aggregate Cold Feed

The aggregate cold feed system is the first major component of the HMA plant. It receives cold (unheated) aggregate from storage and moves it to the dryer. It is again worth noting that none of the plants equipment or control automation can detect or correct inconsistencies in gradation or aggregate quality. Cold feeds can only meter and proportion the supplied aggregate to a given accuracy and consistency. Therefore, quality control of the aggregate production is required to ensure delivery of a uniform, consistent aggregate quality and gradation.

The aggregate cold feed system generally consists of a series of cold feed bins that are filled, or charged, by front end loaders which obtain the aggregates from stockpiles (Figure 4.12). In some cases, large stockpiles separated by bulkheads are built over tunnels, or where belts transport the material, or large storage bunkers or bins are used.

Aggregate from each stockpile should be placed into an individual cold bin. The bins should be kept full enough to ensure a uniform flow through the feeder. Uniform cold feeding is vital to ensure continuous, uniform operations throughout the plant. When cold bins are charged, care should be exercised to minimize segregation and degradation of the aggregate.

Cold feed bins should not be overcharged (heaped) to the point where materials can overflow from one bin to another. Vertical dividers placed between the bins ensure that this does not occur, even when bins are filled to capacity. These bin dividers should not be installed on the face where the loader feeds the bin.

Gates located at the bottom of the bins feed controlled amounts of the different aggregates onto the conveyor carrying them to the dryer. Feeder controls regulate the amount of aggregate

Figure 4.12 Aggregate Cold Feed Bins

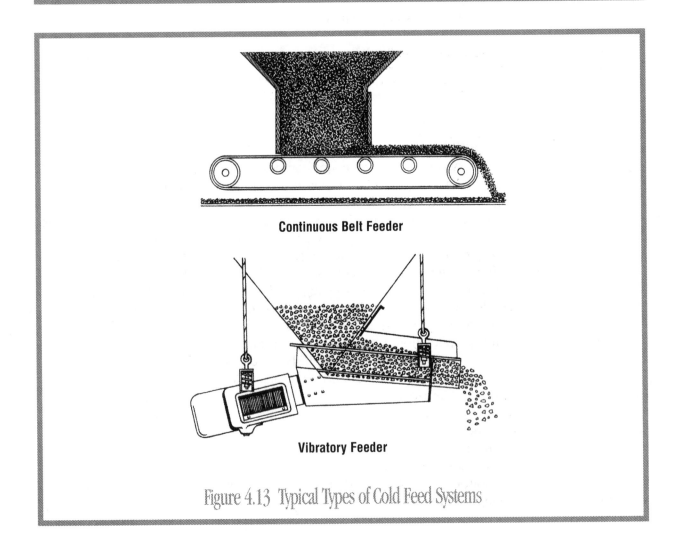

Continuous Belt Feeder

Vibratory Feeder

Figure 4.13 Typical Types of Cold Feed Systems

flowing from each bin, providing a continuous, uniform flow of a properly graded aggregate blend to the plant.

Two of the more common cold feed systems are: (1) continuous belt feeder, and (2) vibratory feeder (Figure 4.13).

Belt feeders provide very good control of the aggregate flow from the individual cold feeds. Material flow is controlled by a combination of belt speed and gate opening. This control capability makes them ideal for drum mix plants, where aggregate gradation is controlled at the cold feed. They are also commonly used on batch plants.

➤➤ *Proper Cold Feeder Function*

Because a uniform flow of properly sized aggregates is so important to consistent HMA production, a check should be made before and during production to be certain that the feeder system is functioning properly. The following conditions are important for maintaining uniform flow and consistency:

- Correct sizes of aggregates in stockpiles and cold bins
- No segregation of aggregates
- No intermixing of aggregate stocks
- Accurately calibrated, set and secured feeder gates
- No obstructions in feeder gates or in cold bins
- Correct speed control settings

Modern HMA facilities have built-in automatic calibration systems. These systems vary widely between plant manufacturers, and therefore, any detailed discussion is beyond the scope of this manual. The manufacturer provides the plant owner with detailed plant calibration instructions, and the plant operator should follow through with these procedures to ensure product quality.

Aggregate Drying and Heating

From the cold bins, aggregates are delivered to the dryer drum. The dryer accomplishes two things. It dries the aggregates, and raises the temperature of the aggregate to the required temperature

The dryer is a revolving cylinder ranging from about 1.5 to 3 meters (5 to 10 feet) in diameter and 6 to 12 meters (20 to 40 feet) in length. It includes an oil or gas burner with a blower fan to provide the primary air for combustion of the fuel (Figure 4.14). An exhaust fan creates a draft through the dryer. The exhaust fan is located beyond the dryer, at the end of the dust control equipment (discussed in the next section).

The drum is also equipped with longitudinal troughs or channels, called flights, which lift the aggregate and drop it in veils through the hot gases (Figure 4.15). The slope of the dryer, rotation speed, diameter, length and the arrangement and number of flights combine to determine the time the aggregate will spend in the dryer.

There are two basic types of dryers, and they are named for the relationship between the flow of the aggregate and the flow of the air within the dryer. Regardless of the dryer style, the principles of drying aggregate are the same.

Figure 4.14 Typical Dryer Drum (Courtesy of Astec)

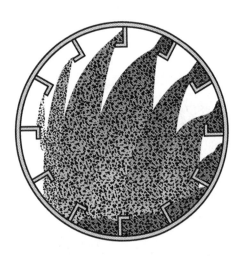

The Double Barrel mixer has a rotating inner drum and a stationary outer drum. This unique design produces optimum flow of materials.

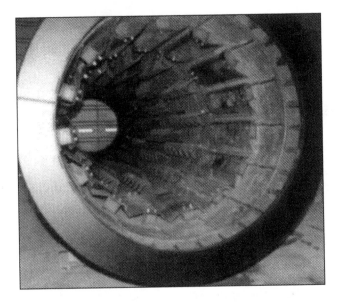

Figure 4.15 Dryer Drum Flights and Function (Courtesy of Astec and Stansteel)

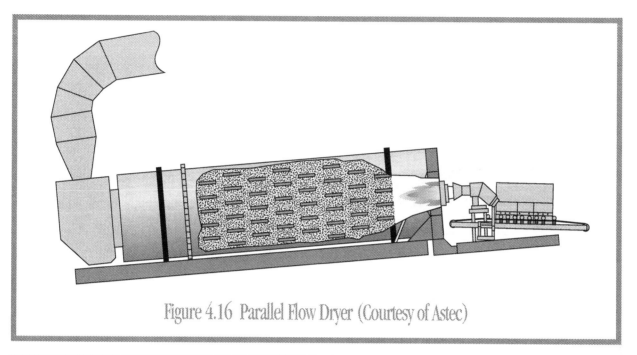

Figure 4.16 Parallel Flow Dryer (Courtesy of Astec)

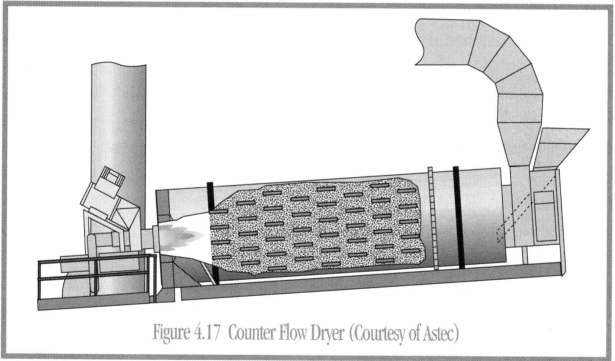

Figure 4.17 Counter Flow Dryer (Courtesy of Astec)

In *parallel flow* dryers (Figure 4.16), the aggregate and the air flow in the same direction. Cold aggregate is introduced into the dryer at the same end as the burner (the source of the air), and the materials flow toward the other end of the dryer.

In *counter flow* dryers (Figure 4.17), the aggregate and air flow in opposite directions, counter to each other. The aggregate is introduced into the dryer at the end opposite the burner and is carried through the drum against the airflow.

Typically, batch plants are equipped with counter flow dryers. When drum mix plants were first developed, they used parallel flow dryers. However, counter flow drum mix plants have

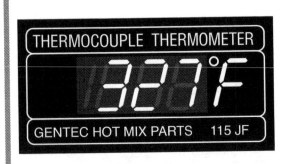

Indicating Pyrometer   **Recording Pyrometer**

Figure 4.18 Types of Pryometers (Courtesy of GenTec and Gencor)

been developed and are the type produced today. Drum mixer styles and configurations are discussed later.

Because of fuel consumption, aggregate drying is the most expensive operation in mix production. Aggregate drying also controls the production rate of the entire facility. HMA cannot be produced any faster than the aggregate can be dried and heated.

Generally, a dryer's capacity is rated for heating and drying aggregate with a given moisture content (typically 5 percent). If the aggregate moisture content is higher, the quantity of aggregate being fed to the dryer must be reduced in order to dry the aggregate properly. Consequently, there is a drop in the dryer's production. Plant elevation, product temperature and aggregate type will also affect dryer performance.

For efficient dryer operation, the air that is combined with the fuel for combustion must be in balance with the amount of fuel being fed into the burner. The exhaust fan creates the draft of air that carries the heat through the dryer and removes the moisture. Imbalance among these elements causes serious problems. When using fuel oil, the lack of sufficient air or excess flow of fuel oil can lead to incomplete combustion of the fuel. The unburned fuel leaves an oily coating on the aggregate particles and can adversely affect the finished mixture.

A quick way to check for oil coating on the aggregate is to place a shovel full of the aggregate being discharged from the dryer into a bucket of water. An oily film will float to the surface. A slight film is not a concern; a heavy film on the surface of the water demonstrates a need for immediate attention to the dryer.

Too much air from the burner can cause a backpressure within the drum. This creates a "puff-back" of exhaust at the burner end of the drum, indicating that draft air velocity is insufficient to accommodate the air pressure created by the burner blower. In such a case, either the resistance to draft air must be reduced or burner air pressure decreased.

Dryers with natural gas or liquid petroleum burners rarely develop combustion problems. However, imbalances among gas pressure, combustion air and draft may still occur.

➤➤ *Temperature Control*

Proper aggregate temperature is essential to mix temperature control. The asphalt added to the aggregate during mixing assumes the aggregates temperature almost immediately. Excessively heated aggregate during mixing can cause accelerated hardening of the asphalt during its life. Underheated aggregate is difficult to coat thoroughly and the resulting mix is difficult to place on the roadway.

A temperature-measuring device called a pyrometer is used to monitor aggregate temperature as the material leaves the dryer. There are two types: (1) indicating pyrometer, and (2) recording pyrometer (Fig. 4.18). The recording head of a recording pyrometer is usually located in the plant control room, while the sensing element is at the discharge chute of the dryer. The indicating pyrometer can be located at the discharge chute of the dryer.

A good temperature-indicating device helps improve plant operation by providing:

- Accurate temperature records
- Indications of temperature fluctuations that suggest lack of control and uniformity in drying and heating operations

Both types of pyrometers are quite similar in operation. In each, the sensing element, which is a shielded thermocouple, protrudes into the main hot aggregate stream in the discharge chute of the dryer.

Pyrometers are sensitive instruments that measure the very small electrical current induced by the heat of the aggregate passing over the sensing element. The head (indicating element) of the device must be completely shielded from heat and plant vibrations. Usually, it is located several feet away from the dryer and is connected to its sensing element by wires. Any change in the connecting wire's length, gauge, splices, or couplings affects the calibration of the device, so that it must be recalibrated after any changes.

The major difference between recording pyrometers and indicating pyrometers is that indicating pyrometers give a dial or digital reading while recording pyrometers record aggregate temperatures on paper in graph form, thus providing a permanent record.

The accuracy of a pyrometer can be checked by inserting the device's sensing element and an accurately calibrated thermometer in a hot oil or asphalt bath. With caution for the baths flash point, the bath is slowly heated above the temperature expected of the dried aggregate and the readings of the two instruments are compared.

Another means of checking a temperature-indicating device is by taking several shovelfuls of hot aggregate from the dryer discharge chute and dumping them in a pile on the ground. Another shovelful is then placed, shovel and all, on top of the pile. The pile keeps the shovelful of aggregate hot while its temperature is taken. Inserting the entire stem of an armored thermometer into the aggregate in the shovel will give a temperature reading that can be compared to the reading on the pyrometer. Several thermometer readings may be necessary to get accurate temperature data.

►►*Moisture Check*

Quick checks for moisture in the hot aggregate can be made at the same time as temperature indicator checks. Quick moisture checks are useful in determining whether more precise laboratory moisture tests should be run.

To make a quick moisture check, a pile of hot aggregate from the dryer discharge is built up and a shovelful of aggregate placed on top of it. The shovelful of aggregate is then checked as follows:

1) The aggregate is observed for escaping steam or damp spots. These are signs of incomplete drying or porous aggregate releasing internal moisture, which may or may not be detrimental. This type of visual check becomes more accurate as the observer becomes more familiar with the aggregate being used.

2) A dry, clean mirror, shiny spatula, or other reflective item which is at normal ambient temperature or colder, is passed over the aggregate slowly and at a steady height. The amount of moisture that condenses on the reflective surface is observed. Excessive moisture can be detected fairly consistently with some practice.

Emission Control Systems

The air flowing through the dryer carries with it exhaust gases and a small amount of the aggregate dust particles. The amount of airborne dust is a function of the size and weight of the material being dried, the velocity of the air in the dryer and, in the case of some drum mixers, the presence of asphalt binder to coat the smaller particles and keep them in the drum and out of the exhaust gas stream.

Because some fine material becomes airborne, emission control equipment must be present to capture this dust before it is discharged into the atmosphere. These emissions must not exceed the various state and federal air pollution limits.

Airflow inside a drum is primarily a function of the drum's diameter. Since the generally accepted industry standard for velocity of air through a drum is about 1,000 feet per minute, the exhaust fan must be sized to ensure that adequate airflow is provided. The velocity of the exhaust gas is a primary factor affecting the potential amount of airborne material for an aggregate with a given size distribution. The amount of airborne dust increases in proportion with the square of the exhaust gas velocity.

One other factor that affects the amount of dust entering the pollution control system in a drum mixer is the location of the asphalt binder inlet. The closer the binder inlet is to the flame, the larger the quantity of airborne dust that is captured by the asphalt binder and prevented from being picked up by the dust collector system. However, moving the asphalt binder inlet closer to the flame produces two potential problems. First, the asphalt binder may be oxidized or stripped of its low volatile components, which produce blue smoke in the stack. Second, the aggregate is not dried as well before being coated. The facility operator must optimize the location of the asphalt binder inlet to minimize the amount of dust and smoke in the pollution control system while at the same time meeting the HMA moisture content requirements of the specifying agency.

The emission control system in most HMA facilities generally consists of primary and secondary dust collectors. The dust collectors are situated at the end of the dryer and filter the air that enters at the burner and exits at the exhaust fan (Figure 4.19).

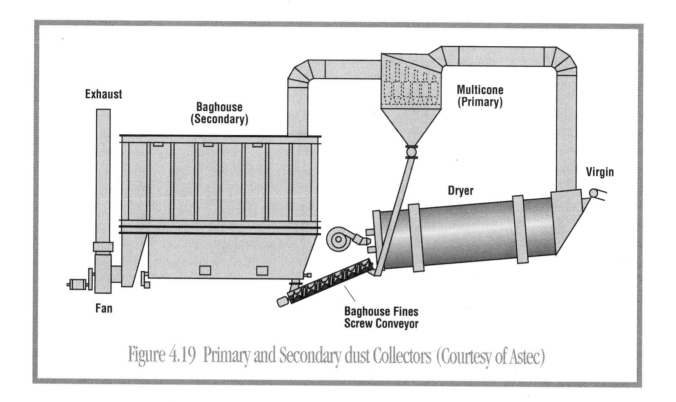

Figure 4.19 Primary and Secondary dust Collectors (Courtesy of Astec)

▶▶ Primary Collectors

The purpose of the primary collector is to collect and remove the larger dust particles contained in the exhaust gas stream. The typical primary collectors are the knockout box and the cyclone collector.

Knockout Box The knockout box is the simplest type of primary collector (Figure 4.20). The exhaust gas flows through an expanded chamber, causing the air speed to decrease. The chamber also contains plates to cause a change in direction of airflow. The speed reduction and direction changes cause the dust particles to drop out of the air stream to the bottom of the box, where they are removed for reuse or disposal.

Cyclone Dust Collector Cyclone collectors are more efficient than knockout boxes, and operate on the principle of centrifugal separation. The exhaust stream cycles around inside the collector (Figure 4.21), and particles hit the outside wall and drop to the bottom of the cyclone. Speed and directional changes also assist, as the exhaust is discharged through the top of the collector. The fines collected at the bottom of the cyclone are picked up by a dust return auger and may be returned to the plant or wasted.

Several styles of cyclone collectors are available. Styles include a large, single cyclone; configurations of multiple, smaller cyclones; and multi-tube, multi-cyclone collectors. In the past, the cyclone collector has been the most common type used, and multiple cyclone configurations were often efficient enough to serve as the sole dust collector, especially in rural areas. However, with the advent of more stringent pollution control laws and the subsequent development of highly effective secondary collectors, knockout boxes or single cyclones are most commonly used as primary collectors in conjunction with a secondary collector.

Figure 4.20 Knockout Box (Primary Dust Collector)

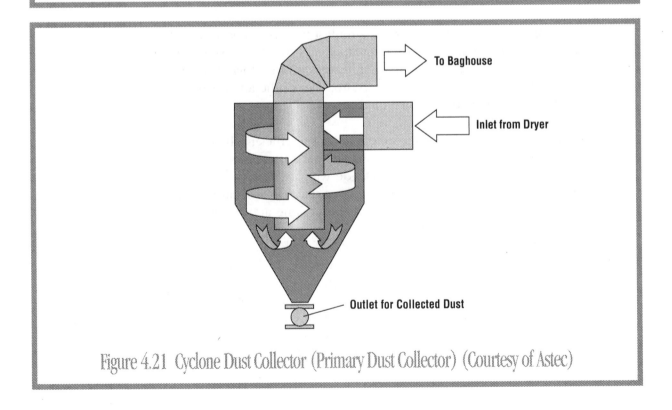

To Baghouse

Inlet from Dryer

Outlet for Collected Dust

Figure 4.21 Cyclone Dust Collector (Primary Dust Collector) (Courtesy of Astec)

Figure 4.22 Baghouse (Secondary Dust Collector) (Courtesy of GenTec)

➤➤ Secondary Collectors

The purpose of the secondary collector is to filter out the finest dust particles. Two types of secondary collectors in common use are the baghouse, and the wet scrubber.

Baghouses A baghouse is a large metal housing containing hundreds of heat resistant filter fabric bags. A typical unit may contain as many as 800 bags. It operates on the principle similar to a vacuum cleaner. The dust-laden exhaust gases are pulled through filter bags supported on long wire cages (Figure 4.22). The dust is trapped on the dirty air side of the bag as the air passes through the filter cloth to the clean air side, thus cleaning the dust from the exhaust stream. When properly operated, baghouses can be very efficient, removing 99 percent of the dust from the dryer exhaust.

Periodically during operation, the bags must be cleaned. This is accomplished by one of several methods. The most common method is to reverse the airflow though the bags with pulsating jets.

The baghouse is divided into sections. While the exhaust gases are diverted from one section to another, the bags in the unused section are cleaned. The dust removed from the bags drops to the bottom of the baghouse and is removed to be wasted or reincorporated into the mix.

Wet Scrubber The purpose of a wet scrubber is to entrap dust particles in water droplets and remove them from the exhaust gases. There are a number of styles of wet scrubbers available that all work on the same principle. Water is broken up into small droplets and those droplets are brought into direct contact with the dust-laden gases. In the case of the centrifugal wet

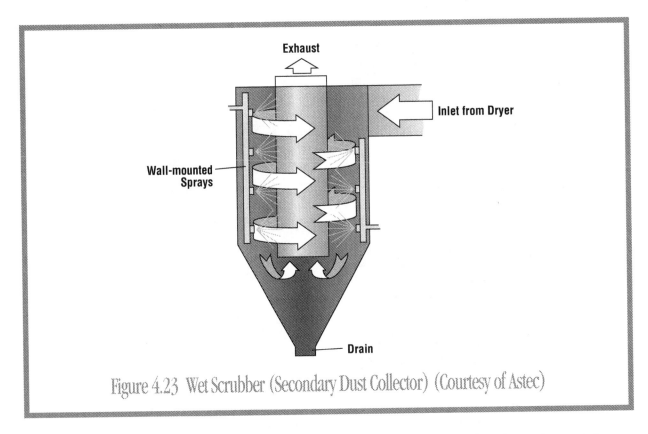

Figure 4.23 Wet Scrubber (Secondary Dust Collector) (Courtesy of Astec)

scrubber (Figure 4.23), gases from the dryer are introduced into a chamber through one inlet while water is sprayed into the chamber from nozzles around the periphery.

Wet scrubbers require a settling basin to receive the scrubber water. The settling basin is designed so that the fines settle out as the water passes through a series of dams and spillways. After the fines settle out, clean water is available for reuse in the system.

Several drawbacks to the wet scrubbers are:

- The dust removed from the exhaust gases is not available for reincorporation into the mix.
- Proper operation requires a large source of water.
- Cleanup and proper disposal of the settlement require additional efforts.

Hot Mix Asphalt Storage

To prevent plant shutdowns due to temporary interruptions of paving operations or shortages of trucks to haul material from the plant to the paving site, most asphalt plants are equipped with *storage silos* or *surge bins* for temporary storage of HMA. Newly made HMA is deposited by conveyor or hot elevator into the top of the silo or bin and is discharged into trucks from the bottom. Insulated silos, sometimes called storage bins as well as storage silos (Figure 4.24), can store hot mix up to 24 hours with no significant loss of heat or quality. Capacities range as high as several hundred tons. Non-insulated storage structures, generally called surge bins (Figure 4.25), are usually smaller and can store hot mix only for short periods of time.

Storage silos work well if certain precautions are followed, but they can cause segregation of the mix if not used properly. A batcher should be used in conjunction with the silo to keep segregation to a minimum. The mix is conveyed into a batcher at the top of the silo. Once a prede-

Figure 4.24 Storage Silos (Gencor)   Figure 4.25 Surge Bin (Courtesy of Gencor)

termined amount is deposited in the batcher, it releases the mix into the silo in a single mass. (Figure 4.26)

During continuous operation, the mix level in the storage silo should be maintained between one third and two thirds of the silos capacity.

➤➤ *Weighing and Hauling*

Hot mix asphalt is hauled to paving sites in various types of trucks. Hauling trucks vary by size and type, but uniformity of equipment is very desirable in any asphalt paving operation. The trucks should be inspected carefully before use. For accurate material control, truckloads of HMA must be weighed at the plant.

The quantity of hot mix delivered from plant to paving site can be determined by either of two methods: (1) weighing loaded trucks on scales, or (2) using a plant's automatic recording system (in the case of fully-automated plants). When truck scales are used, they must be of the type that directly indicates the total weight of the truck. They must be horizontal and of sufficient size to weigh all of a truck's axles at one time. The most common type of truck scale used is the beam scale (Figure 4.27).

The accuracy of truck scales must be checked periodically. For this purpose, the contractor loads a truck with some type of material, weighs the loaded truck on the scales, and then weighs it on another set of certified truck scales. Truck scales must also be balanced before use.

During a normal days operation, the scale should be checked frequently to make sure that it is in balance. Mud or other foreign matter left on the platform by the trucks can throw the scale out of balance. If there is very little foreign matter on the platform, the scale can be rebalanced

Figure 4.26 A Batcher Above a Surge Bin

by adjusting the counterweight. If adjusting the counterweight will not balance the scale, the platform will have to be cleaned. If, after the platform is cleaned, the scale still does not balance, plant operations must be stopped until the scale is working properly.

In addition to periodic checks of the scale and platform, each truck must be randomly tared (weighed when empty). A permanent record of the tare weight must be maintained in the scale house and, when required, in the truck. Electronic automated printout weigh tickets are now accepted by a number of states and other agencies. These tickets usually will contain the gross, tare and net weights.

Figure 4.27 Truck Scale (Courtesy of Stansteel)

Quality Control and Observation of HMA Production

Quality control (QC) of the produced HMA is an important part of plant production. Compliance of the mixture with specification requirements, performance of the pavement and proper compensation to the producer/contractor depend to a great extent on properly executed QC procedures. These procedures include not only sampling and testing, but also adjustment of the mix and the process to keep the produced mix in compliance with the specifications.

While either the agency or its representative can perform QC sampling, testing and inspection, a majority of the specifications require that the contractor perform QC procedures. Chapter seven, Quality Control and Acceptance of Hot Mix Asphalt, explains these procedures.

In conjunction with QC sampling and testing, visual observation by contractor and agency personnel can detect problems and defects in the mix. This should be done as the mix is discharged from the plant and loaded into the trucks and while the mix is in the trucks.

Temperature control of the HMA is important in controlling quality. A visual observation can often detect whether or not the temperature of a load of mix is within the proper range. Blue smoke rising from a truckload of mix is usually an indication of overheating. If the mix temperature is too low, the mix may appear sluggish as it is deposited in the truck, and it may show a non-uniform distribution of asphalt. An abnormally high peak in a truckload may also indicate underheating.

Normally, the temperature of the mix is taken in the truck. The most common way to determine the temperature is with a dial and armored-stem thermometer. The stem should be inserted sufficiently deep into the mixture—at least 150mm (6 inches)—and the HMA should be in direct contact with the stem. A gun-type infrared thermal meter, which measures reflective heat from a surface, may also be used. Because it detects only surface heat, the meters temperature readings may not be accurate for material in the middle of the load. To overcome this, the instrument should be aimed at the stream of mix as it leaves the discharge gate of the mixer or storage silo. Infrared thermal meters give fast readings, but they should be used with caution in determining specification compliance, and they should be calibrated frequently.

An abnormally high peak in the load can also be an indication that the asphalt content of the mixture is too low. On the other hand, if the mix slumps (fails to peak properly) in the truck, it may contain excessive asphalt or excessive moisture.

There are many causes of visible nonuniformity in plant produced HMA. Table 4.02 provides information on various deficiencies in mixtures and their possible causes.

The following checklist can be used by plant operation and quality control personnel to determine plant readiness for production of HMA:

HMA PLANT CHECKLIST

Checklist For Material Handling and Storage
- Do aggregates meet specifications?
- Are proper sizes being produced?
- Is aggregate storage satisfactory?
- Are stockpiles separated properly?
- Are stockpiles constructed properly?
- Is stockpiled aggregate handled correctly?
- Is segregation being controlled?
- Is mineral filler or hydrated lime being kept dry?

Checklist for Cold Feed
- Does cold feed set-up comply with specifications?
- Do cold feed bins contain proper size aggregates?
- Are cold feed bins charged properly?
- Do cold aggregate feeders perform satisfactorily?
- Are cold aggregate feeders calibrated?
- Are cold aggregate feeder gates set correctly?
- Are all cold aggregates feeding continuously?

Table 4.02 Possible Causes of Deficiencies in Hot Mix Asphalt

Types of deficiencies that may be encountered in producing hot mix asphalt.

Aggregates Too Wet	Inadequate Bunker Separation	Aggregate Feed Gates Not Properly Set	Oven-Rated Dryer Capacity	Dryer Set Too Steep	Improper Dryer Operation	Temp. Indicator Out of Adjustment	Aggregate Temperature Too High	Worn Out Screens	Faulty Screen Operation	Bin Overflows Not Functioning	Leaky Bins	Segregation of Aggregates in Bins	Carryover in Bins Due to Overloading Screens	Aggregate Scales Out of Adjustment	Improper Weighing	Feed of Mineral Filler Not Uniform	Insufficient Aggregates in Hot Bins	Improper Weighing Sequence	Insufficient Asphalt	Too Much Asphalt	Faulty Distribution of Asphalt to Aggregates	Asphalt Scales Out of Adjustment	Asphalt Meter Out of Adjustment	Undersize or Oversize Batch	Mixing Time Not Proper	Improperly Set or Worn Paddles	Faulty Dump Gate	Asphalt and Aggregate Feed Not Synchronized	Occasional Dust Shakedown in Bins	Irregular Plant Operation	Faulty Sampling	Types of deficiencies
	A													B	B				A	A	A	B	C	B	B	B		C			A	Asphalt Content Does Not Check Job Mix Formula
A	A							B	B	B	B	A	A	B	B	B	A					B				B	B	C	B		A	Aggregate Gradation Does Not Check Job Mix Formula
A	A								B	B	B	A	A	B	B	B	A					B	B					C	B		A	Excessive Fines in Mix
A			A	A	A	A	A																							A		Uniform Temperature Difficult to Maintain
											B			B	B							B										Truck Weights Do Not Check Batch Weights
														B	B				A	A	B	C	B			B		C				Free Asphalt on Mix in Truck
																		B									B					Free Dust on Mix in Truck
A			A	A	A														A		A	B	C	B		B		C			A	Large Aggregate Uncoated
											B	B	A	A	A	B	B	B	A	B	A	B	C			B	B	C	B	B	A	Mixture in Truck Not Uniform
																		B			A				B	B	B				A	Mixture in Truck Fat on One Side
							A												A	A	B	C	B					C			A	Mixture Flattens in Truck
	A			A	A	A																									A	Mixture Burned
A			A	A	A	A			B										A			B	C	B				C			A	Mixture Too Brown or Gray
														B	B	B	A		A	A	B	C	B					C			A	Mixture Too Fat
				A	A	A																									A	Mixture Smokes in Truck
A				A	A	A																									A	Mixture Steams in Truck
				A	A	A													A											A	A	Mixture Appears Dull in Truck

A – Applies to Batch and Drum Mix Plants; B – Applies to Batch Plants; C – Applies to Drum Mix Plants

Checklist for Asphalt Heating, Circulating and Temperature of Mixture
- Is asphalt uniformly heated to the temperature specified?
- Have all lines been checked for leaks?
- Is the specified temperature of the mixture and its components being maintained?

Checklist for Drum Mix Plant
- Have aggregate feeds been calibrated?
- Has asphalt feed been calibrated?
- Are aggregate and asphalt feeds interlocked?
- Are all plant parts in good condition and adjustment?
- Is the asphalt at the proper temperature when introduced into the drum?

Checklist for Batch Plant

- Do scales comply with specifications?
- Have scales been calibrated?
- Have scales been checked for tolerance?
- Does asphalt bucket tare properly?
- Does weigh box hang free?
- Are mixer parts in good condition and adjustment?
- Is proper size batch being mixed?
- Is bin withdrawal in proper sequence?
- Is asphalt distribution uniform along the pugmill?
- Are aggregates and asphalt at proper temperatures when introduced into the weighing receptacles?
- Do any valves or gates leak?
- Is mixing time adequate?
- Are weight points set properly for batch weights?
- Are mixer shafts revolving at proper speed?
- Are the screen capacities sufficient to handle the maximum feed from the dryer?
- Are screens clean?
- Are screens worn or broken?
- Is the carry-over irregular or excessive?
- Are hot bin partitions sound?
- Are overflow chutes free-flowing?
- Is bin balance being maintained?
- Is access for sampling adequate?

Checklist for Dryer and Dust Collector

- Do dryer and dust collector comply with specifications?
- Is the aggregate properly dried?
- Are the aggregates at the proper temperature?
- Are dryer components in balance?
- Is dryer in balance with other plant components?
- Is the heat-indicating device installed correctly?
- Has the heat-indicating device been checked for accuracy?
- Is dust collector in balance with dryer?
- Are collected fines from the dust collector wasted, or fed back uniformly in the desired amount?

Checklist for Storage Silos

- Does the silo contain a batcher?
- Are baffles or other devices to prevent segregation working properly?
- Is the silo discharge opening properly configured to prevent segregation?
- Does the discharge gate open and close efficiently?

Checklist for Sampling and Testing

- Are sufficient samples being taken to comply with the sampling plan?
- Are samples representative of the material?
- Are all tests being conducted properly?
- Are test results available soon enough to be effective?

Checklist for Records
- Are records complete and up-to-date?

Checklist for Miscellaneous Responsibilities
- Have truck beds been inspected?
- Are truck beds drained after spraying?
- Do trucks meet specification requirements?
- Are trucks equipped with tarpaulins or covers?
- Is the mix of uniform appearance?
- Is the general appearance of the mix satisfactory?
- Is the temperature of the mix uniform and satisfactory?
- Does the mix satisfy the placing requirements?
- Have all personnel been properly instructed?
- Are safety measures being observed?

A daily summary report of plant activities also should be kept. There should be a summary of the results of all tests performed during the day and a tabulation of the amounts of material received and used.

Safety

Personnel working at a hot mix asphalt plant must always be safety-conscious and on the alert for potential dangers to personnel and property. Safety considerations cannot be overemphasized.

Dust is particularly hazardous. It is not only a threat to lungs and eyes, but it may contribute to poor visibility, especially when trucks, front-end loaders, or other equipment are working around the stockpiles or cold bins. Reduced visibility in work traffic is a prime cause of accidents.

Noise can be a double hazard also. It is harmful to hearing and can distract workers' awareness of moving equipment or other dangers.

Moving belts that transport aggregates should be a constant concern, as should belts to motors and sprocket and chain drives. All pulleys and belts and drive mechanisms should be covered or otherwise protected. Loose clothing that can get caught in machinery should never be worn at an asphalt plant.

Good housekeeping is essential for plant safety. The plant and yard should be kept free of loose wire or lines, pipes, hoses or other obstacles. Such hazards as high voltage lines, field connections, or wet ground surfaces should be noted. Any loose connections, frayed insulation, or improperly grounded equipment should be reported immediately.

Plant workers should not work on stockpiles while the plant is in operation. No one should walk or stand on the stockpiles or on the bunkers over the feeder gate openings. Many persons have been pulled down into the material and buried alive so quickly that they had no warning.

Burner flames and high temperatures around plant dryers are obvious hazards. Installing control valves that can be operated from a safe distance on all fuel lines helps reduce the danger. Flame safety devices also should be installed on all fuel lines. Smoking should not be permitted near asphalt or fuel storage tanks. Frequent checks should be made for leaks in oil heating lines and steam lines or jacketing on the asphalt distribution lines. Safety valves should be installed in all steam lines, and they must be in working order. Screens, barrier guards, and shields should be installed as protection from steam, hot asphalt, hot surfaces, and similar dangers.

When handling *heated asphalt*, chemical goggles and a face shield should be used. All shirt collars should be worn closed and cuffs buttoned at the wrist. Gloves with gauntlets that extend up the arm should be worn loosely so that they can be flipped off easily if covered with hot asphalt. Pants without cuffs should extend over boot tops.

Extreme care should be exercised when climbing around the screen deck observing the screens and hot bins or collecting hot bin samples. There should be covered or protected ladders or stairways to provide safe access to all parts of the plant. All stairs and platforms should be provided with secure handrails. All workers around the plant site should always wear a hard hat when not under cover.

Truck traffic patterns should be planned with both safety and convenience in mind. Trucks entering the plant to pick up a load of hot mix should not have to cross the path of loaded trucks leaving the plant. In addition, trucks should not have to back up.

Batch Plants

Batch plants get their name from the fact that during operation they produce HMA in batches, one batch at a time, one after the other. The size of a batch varies according to the capacity of the plants pugmill (the mixing chamber where aggregate and asphalt are blended together). Batch sizes can vary from 1,800 to 7,200 kilograms (4,000 to 16,000 pounds).

➤➤ History of Batch Plants

The basic operations of HMA production – drying, screening, proportioning, and mixing – were first combined into an asphalt plant around 1870. Early plants, although crude by today's standards, formed the basis for HMA production during the nineteenth century.

By 1900, plants had been improved to include aggregate bins, cold elevators, rotary dryers, hot elevators, asphalt tanks, and mixing platforms. Mixing platforms featured an aggregate measuring box, an asphalt bucket, and a pugmill mounted high enough to allow horse-drawn carts to pass underneath.

By 1930, plants were producing 800 to 1,000 tons per eight-hour day. In the 1930s and 1940s, better cold feed systems resulted from the introduction of conveyor belts and the improvement of gates and feeders. Larger dryers came into common use. Cyclone dust collectors, springless scales, automatic electronic weighing systems, time locks on the mixing cycles, and temperature recorders appeared.

Larger, higher-capacity plants dominated development in the 1950s. Adoption of automatic burner controls and automation of the proportioning and cycling functions also came into use early in this period.

The 1960s saw a proliferation of automatic control systems, with full automation of the proportioning and mixing process, as well as automated burner control systems.

The most significant developments of the 1970s were the emergence of computerized plant control systems, the use of load cells, and improvements in noise and dust control, stemming from increasing health and safety regulations.

The 1980s and 1990s brought continuing refinement of computer controls and improved accuracy and ruggedness of load cells, as well as efficient computerized calibration controls. With the improved computerized controls, some batch plants are able to operate without the

Figure 4.28 Batch Plant (Courtesy of Stansteel)

use of screens and separate bins. The tighter controls allow all the dried and heated aggregate to be placed in one bin and batched from that single bin. This offers the advantage of being able to switch mixture types without resetting the screen deck. The 1990s also brought more efficient dust collection and return systems. This enables the industry to better control the amount of material passing the 0.075 mm (No. 200) sieve in HMA mixtures.

Throughout all the changes to the batch plant, the fundamental process – drying, screening, proportioning, and mixing – has remained. The basic design of the equipment for performing these operations has changed little since 1940; only the size and capacity have increased.

➤➤ *General*

As discussed earlier, the basic operations of both batch and drum mix plants are similar except for the aggregate and asphalt mixing procedures. Material storage and handling, aggregate cold feeds, aggregate heating and drying, emissions control, and HMA storage and handling are all quite similar for batch and drum mix plants. Those similar processes are covered in previous sections in this chapter. This section will focus on the batch plant (Figure 4.28) procedures after the aggregate has been dried and heated in the dryer.

▶▶Screening and Storage of Hot Aggregate

After the aggregate has been heated and dried, it is deposited from the dryer into the hot elevator. The hot elevator is an enclosed bucket conveyor that carries the aggregate to the top of the "batching tower." At this point, the aggregate is dropped onto the screening unit. Here, the hot aggregate passes over a series of screens that separate it into various-size fractions and deposit those fractions in hot bins.

Screening Unit The screening unit includes a set of several different-size vibrating screens inside a large housing (Figure 4.29). The screens separate the aggregate into specific sizes. The first screen is a scalping screen that removes oversized aggregate. This is followed by two or three screens, decreasing in size from top to bottom. The sizes of the screens to be used will depend on the particular plant and the gradation of the aggregate for the mix to be produced (Figure 4.30). The screens are designed to allow the finest particles to drop completely through to the first hot bin, and larger particles move along the screens to be deposited into the other bins.

To properly screen the aggregate, the total screen area must be large enough to handle the amount of aggregate delivered. The capacity of the screens must be in balance with the capacity of the dryer and the capacity of the pugmill. When too much material is fed to the screens, particles that should pass through will ride over the screens and drop into a bin designated for a larger size particle. Any misdirection of a finer aggregate into a bin intended to contain the next larger size fraction is called carry-over, and the screens are said to be blinded.

The screens must be clean and in good condition. When screen openings are plugged, carry-over will occur. Similarly, when screens are worn or torn resulting in enlarged openings and holes, oversized material will go into bins intended for smaller-size aggregate.

Excessive carry-over can add to the amount of fine aggregate in the total mix, thus increasing the surface area to be covered with asphalt. If the amount of carry-over is unknown or if it fluctuates, particularly in the No. 2 bin, it can seriously affect the mix composition in both gradation and asphalt content. Excessive carry-over can be detected by a sieve analysis of the contents of the individual hot bins and must be corrected immediately by cleaning the screens or reducing the quantity of material coming from the cold feed, or both. Some carry-over is permitted in normal screening and the permissible amount of carry-over in each bin is usually specified.

The No. 2 bin (intermediate fine aggregate) is the critical bin for carry-over. This is the bin that will receive the finest aggregate in carry-over and which will affect the asphalt demand of the mix the most. Typically, the carry-over in the No. 2 bin should not exceed 10 percent of the material in that bin. Running a sample of the No. 2 bin material over a 2.36 mm (No. 8) sieve will indicate the amount of carry-over.

Small amounts of carry-over are not necessarily detrimental to the proportioning process as long as the amount of carry-over can be consistently maintained and accounted for in the bin percentages used in the mix.

Hot Bins Hot bins are located directly below the screening unit and are used to temporarily store the heated and screened aggregates in the various sizes required. Each bin is an individual compartment, or it can be considered to be a segment of a large compartment divided by partitions. Properly sized hot bins hold enough material of each size when the mixer is operating at full capacity. The hot bins must be tight, free from holes and high enough to prevent intermingling of the aggregates.

Hot bins usually have indicators that tell when the aggregates fall below a certain level. These indicators may be either electronic or mechanical. One such electronic indicator is mounted on the side of the bin (Figure 4.31).

Figure 4.29 Screening Unit (Courtesy of Gencor)

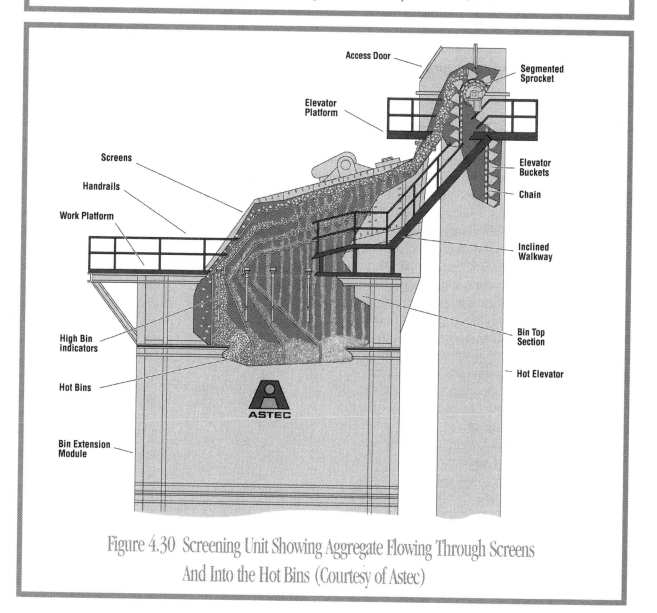

Figure 4.30 Screening Unit Showing Aggregate Flowing Through Screens
And Into the Hot Bins (Courtesy of Astec)

Figure 4.31 An Electronic Low Bin Indicator (Courtesy of Gencor)

Each bin should be equipped with an overflow pipe or an electronic high bin indicator to prevent excess amounts of aggregate from backing up into the other bins. The devices should be set to stop overfilling of the bins. When a bin overfills, the screen above it rides on the aggregate, resulting in a heavy carry-over and possible damage to the screen. Overflow vents should be checked frequently to make sure they are free flowing.

Hot bins are referred to by their total capacity. This can be as low as 18 metric tons (20 tons) for a small, portable batch plant to as large as 272 metric tons (300 tons) for a large, stationary plant. A very common capacity is in the range of 36 to 72 metric tons (40 to 80 tons).

Sometimes the very fine aggregate will hang up in the corners of the fine aggregate bin. When this build-up collapses, it can result in an excessive amount of fines in the mix. This rush of fine materials usually occurs when the aggregate level in the bin is drawn down too low. The solution is to maintain proper aggregate level in the bin. Also, fillet plates welded into the corners of the bin will minimize the build-up of the fines.

Other potential obstacles to a good mix include shortage of material in one bin (and excess in another), worn gates in the bottom of a bin (allowing leakage of aggregate into the weigh hopper), and sweating of the bin walls (caused by condensation of moisture).

Hot bins must not be allowed to run empty. Bin shortages or excesses are corrected by adjusting the cold feed. For example, if the coarse bin is overflowing while the others remain at satisfactory level, the cold bin feed supplying most of the coarse aggregate should be reduced slightly.

It is not good practice to make two cold feed adjustments at once. For example, if the total feed is deficient and one bin is running a little heavy, it is better to adjust the total feed first and then make an adjustment to the feed on the one bin that is running heavy.

If the gate at the bottom of a hot bin is worn and leaking material, it must be repaired or replaced immediately. Leakage from a hot bin can adversely affect gradation of the final mixture.

Sweating occurs when moisture vapor in the aggregate and in the air condense on the hot bin walls. It happens usually at the beginning of the days operations or when the coarse aggregate is not thoroughly dry. Sweating may cause the accumulation of dust that results in excessive surges of fines in the mix.

Neither mineral filler nor dust from the baghouse is run through the hot bins. They should be stored separately in moisture-proof silos and fed directly into the weigh hopper.

►►*Hot Bin Sampling*

Modern hot mix asphalt plants are equipped with devices for sampling hot aggregate from the bins. Most plants utilize a sample container mounted on a rod, which is slid under the hot bin feeder, or gate, to catch the aggregate as it drops. Some plants may have a device to divert the flow of aggregate from the hot bin to a sample container. It is essential that such sample containers be properly located when taking the sample, so that a representative sample of the material in the bins will be collected.

During production, as the aggregate flows over the plant screens, finer particles fall to one side of each bin and coarser particles to the other (Figure 4.32). When material is drawn from the bin by opening a gate at the bottom, the stream consists predominantly of finer material at one end and coarser material at the other. Therefore, the position of the sampling device in the stream of material discharged from a bin determines whether the sample will be composed of a finer portion, a coarser portion, or an accurate representation of the material in the bin (Figure 4.33). This condition is especially critical in the Number 1 (fine) bin, since the material in this bin strongly influences the amount of asphalt required in the mix.

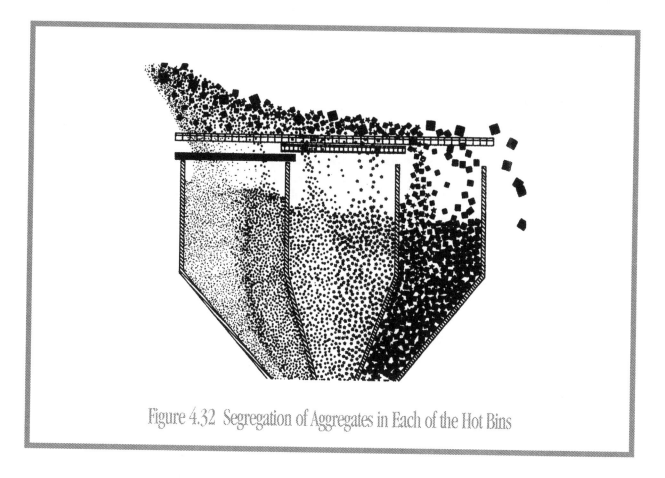

Figure 4.32 Segregation of Aggregates in Each of the Hot Bins

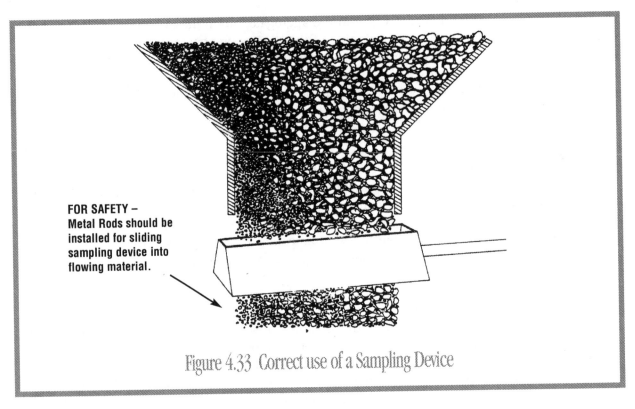

FOR SAFETY –
Metal Rods should be installed for sliding sampling device into flowing material.

Figure 4.33 Correct use of a Sampling Device

Stratification (vertical layering) of sizes in the fine bin can also occur. It may be caused by variations of grading in the stockpiles or by erratic feeding of the cold aggregate. When this form of segregation exists, representative samples cannot be obtained, even when the sampling device is used correctly.

➤➤ Calibration

Normally it is the contractors responsibility to calibrate the HMA plant; however, agency inspectors are often required to observe and be aware of the procedures used to arrive at an aggregate combination that meets the job mix formula.

To produce the desired aggregate combination, the content of each bin must be analyzed. The first step in analyzing the hot bins is to start the plant and the cold feed, the dryer, and the screens. When the plant reaches operating condition, such that it is believed the material in the bins is representative of the proportions established at the cold feed gates, a sample of aggregate is taken from each bin. The aggregate samples are then graded. Once the gradation of material in each hot bin is determined, the exact percentage to be pulled from each bin to meet the design mix can be calculated. This is done by a trial and error method.

➤➤ Drawing Material From The Hot Bins

The aggregate is drawn from each hot bin into the weigh hopper, one bin at a time. The weigh hopper is suspended from scale beams and weighs cumulatively the amounts of aggregate entering it.

The contractor or the producer determines the order in which the bins dump their aggregates into the weigh hopper. Usually, the coarsest aggregate is drawn (or pulled) first, the inter-

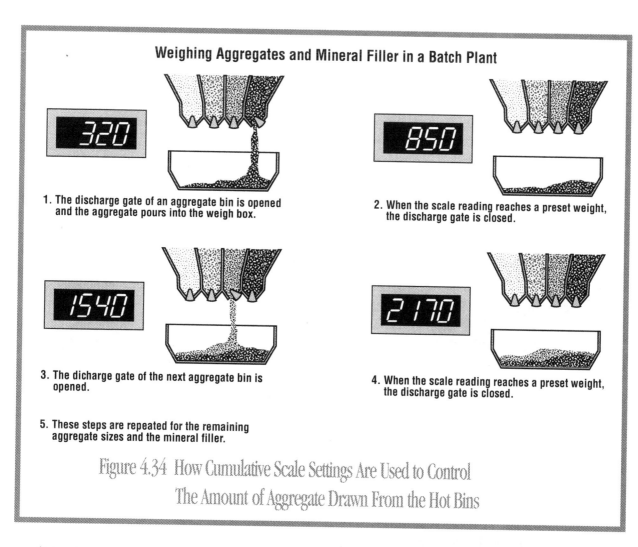

Weighing Aggregates and Mineral Filler in a Batch Plant

1. The discharge gate of an aggregate bin is opened and the aggregate pours into the weigh box.

2. When the scale reading reaches a preset weight, the discharge gate is closed.

3. The dicharge gate of the next aggregate bin is opened.

4. When the scale reading reaches a preset weight, the discharge gate is closed.

5. These steps are repeated for the remaining aggregate sizes and the mineral filler.

Figure 4.34 How Cumulative Scale Settings Are Used to Control The Amount of Aggregate Drawn From the Hot Bins

mediate-size aggregates next, and the finest aggregate last. This sequence is designed to place the fine fractions at the top of the weigh hopper, where they cannot leak out through the gate at the bottom of the weigh hopper. This system also allows the most efficient utilization of the available volume in the weigh hopper, since the finer aggregates will partially penetrate the voids in the coarser aggregates.

When the pulling sequence is determined, the weights to be drawn from the hot bins are set on the scale. This information is normally entered into a computer that controls the opening and closing of the bins to obtain the correct amount of aggregate from each bin. Figure 4.34 illustrates how the cumulative scale settings are used to control the weight of aggregate drawn from each bin.

➤➤ *Introducing The Asphalt*

From the weigh hopper, aggregate is deposited into the plant's pugmill (mixing chamber), where it is blended with the proper proportion of asphalt binder. In a typical plant system, asphalt is weighed separately in weigh buckets—which are enclosed, sealed units— before being introduced into the pugmill. When the weight of asphalt in the bucket reaches a predetermined level, a valve in the delivery line closes to prevent excess asphalt from being discharged into the bucket. The asphalt is then pumped through spray bars into the pugmill (Figure 4.35).

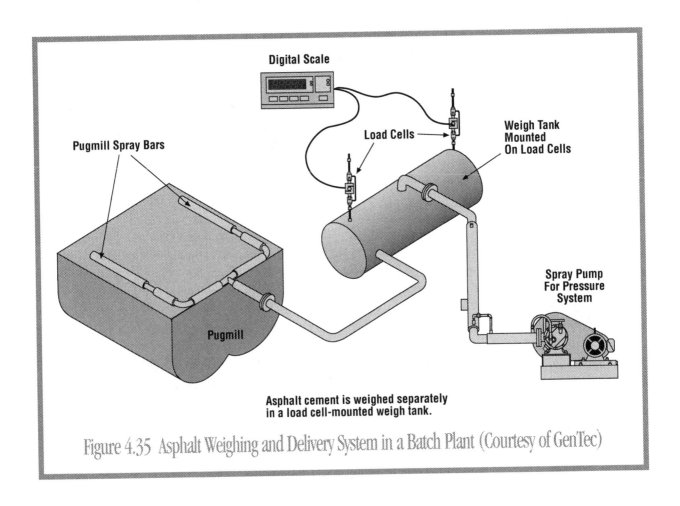

Digital Scale

Pugmill Spray Bars

Load Cells

Weigh Tank Mounted On Load Cells

Spray Pump For Pressure System

Pugmill

Asphalt cement is weighed separately in a load cell-mounted weigh tank.

Figure 4.35 Asphalt Weighing and Delivery System in a Batch Plant (Courtesy of GenTec)

▶▶ Temperature Of Mixtures

Both asphalt and aggregate must be heated before they are combined in the pugmill – the asphalt to make it fluid enough to pump and properly coat the aggregate, and the aggregate to make it dry and hot enough to accept the asphalt and produce a finished mix at the desired temperature.

Asphalt is a thermoplastic material, which means that its viscosity decreases as its temperature increases and vice versa. The relationship between temperature and viscosity, however, may not be the same for different sources or types and grades of asphalt. The temperature of the aggregate controls the temperature of the mixture. However, the aggregate and the asphalt temperature should be essentially the same when they are introduced into the pugmill. Normally, a mixing temperature is specified, based on the characteristics of the asphalt and on factors relating to mixing, placement, and compacting conditions. Another consideration is the temperature required to dry the aggregate sufficiently to obtain a satisfactory mix.

Mixing should be done at the lowest temperature that will provide for complete coating of the aggregate particles and produce a mixture of satisfactory workability. With the introduction of the performance graded (PG) asphalt binders, the selection of the correct mixing temperature has become more complex. The mixing temperature of the binder, as determined from the laboratory viscosity tests and the temperature-viscosity chart, can be used as a point of reference. In some cases, for neat (unmodified) asphalts, this may be the proper mixing temperature or, at least, close to it. The binder producer's recommendation for mixing temperature is considered accurate and should be used if at all possible. It should be understood that two asphalt binders with the same performance grade could have different optimum mixing temperatures, especially if they are produced by different methods. And they most certainly will have different mixing temperatures if one is a modified asphalt and the other is not.

➤➤ Pugmill Mixing

Once the batching of aggregate and asphalt is completed, the aggregate is transferred to a mixing chamber, called a *pugmill*. The pugmill is situated immediately below the weigh hopper.

Virtually all modern plants use a twin-shaft pugmill, which consists of a lined mixing chamber with two horizontal shafts, on which several cross arms are mounted. At each end of the cross arms is a metal plate, commonly called a mixing paddle. These paddles must be adjusted to avoid dead zones in the pugmill. Dead zones are areas where the materials are not properly and thoroughly mixed. To avoid this situation, the paddles should be adjusted so that the clearance between the paddle tips and the mixer liner is less than one half the maximum size aggregate used in the mix. Worn or broken paddles should be replaced as soon as possible.

If the pugmill is over-filled or under-filled, non-uniform mixing will occur. With too little material in the pugmill, the paddles are not able to adequately mix the material (Figure 4.36). If the pugmill is overloaded, part of the material will tend to float on the top of the batch and not be thoroughly mixed (Figure 4.37). These situations can be avoided by following the manufacturers recommended batch rating. Normally the manufacturers rating is based on a percentage of the pugmills live zone (Figure 4.38). This live zone is the net volume of the inside of the pugmill below a line extending across the top arc of the mixing paddles. The volume of the shafts, cross arms, paddles, and liner is not included in this volume. In most cases, the maximum operating efficiency of a pugmill is achieved when the paddle tips are barely visible at the surface of the material during mixing.

The complete mixing cycle, during which asphalt, aggregates, and mineral filler are blended to produce HMA in the pugmill, is illustrated in Figure 4.39. The length of time between the opening of the weigh box (hopper) gate (Step 1 in the figure) and the opening of the pugmill discharge gate (Step 4) is referred to as the batch mixing time. The batch mixing time must be long enough to produce a homogenous mixture of evenly distributed and uniformly coated aggregate particles. However, if the mixing time is too long, the lengthy exposure of the thin asphalt film to the high aggregate temperature in the presence of air can adversely affect the asphalt and reduce the durability of the mix. To monitor batch mixing time, most job specifications require the use of some type of timing device.

Mixing time may be set within specification limits for each mix in any given plant by the procedure described in AASHTO T195 or ASTM Designation D 2489, Determining Degree of Particle Coating of Bituminous-Aggregate Mixtures. This system bases the degree of mixing on the percentage of coarse particles that are 100 percent coated with asphalt and correlates it with mixing time.

Coarse particles only are used in this procedure because they are the last to be coated in the mixing process. Typical minimum percentages required for specification compliance are 90 percent fully coated for base mixes and 95 percent fully coated for surface mixes. The least time needed for the pugmill to produce a batch meeting the minimum coating requirement is set as the minimum mixing time.

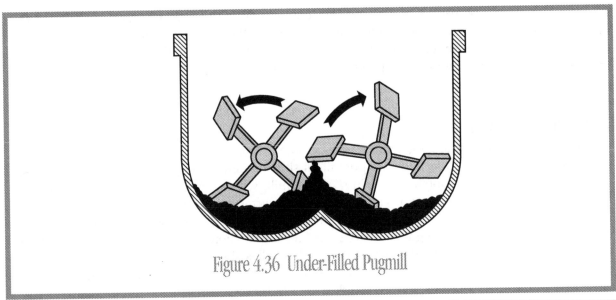

Figure 4.36 Under-Filled Pugmill

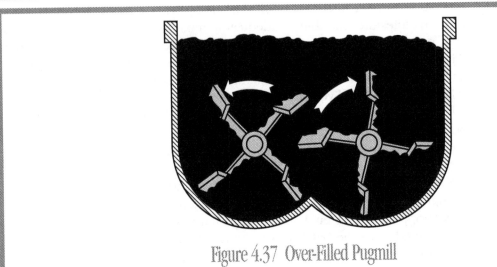

Figure 4.37 Over-Filled Pugmill

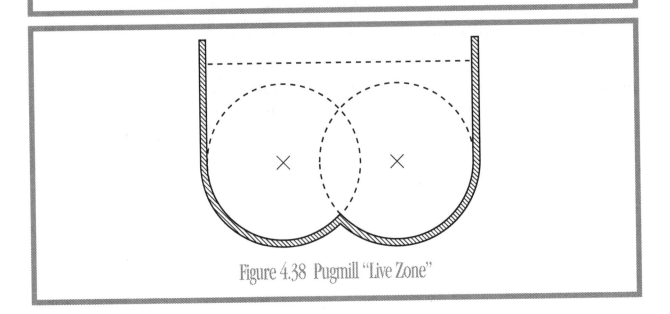

Figure 4.38 Pugmill "Live Zone"

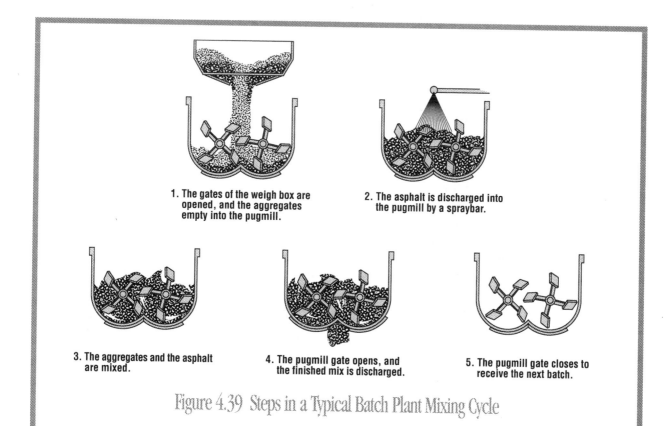

1. The gates of the weigh box are opened, and the aggregates empty into the pugmill.

2. The asphalt is discharged into the pugmill by a spraybar.

3. The aggregates and the asphalt are mixed.

4. The pugmill gate opens, and the finished mix is discharged.

5. The pugmill gate closes to receive the next batch.

Figure 4.39 Steps in a Typical Batch Plant Mixing Cycle

➤➤ *Batch Plant Automation*

Computer automation has many control and testing features. The first is self-learning logic that watches the target values and actual weights and automatically adjusts for changes in actual-to-target. It averages the corrections over a number of pulls, or gives the operator the option to force a value on a special screen.

Secondly, it will warn when an individual material pull is out of tolerance, or if the combined pulls are out of tolerance on the whole. It will give an opportunity to abort a batch, or make adjustments in individual materials or the asphalt to bring it back into specification.

Third, while readying for the next batch, it will print every individual ingredient weight, the time, date, and cycle times, note the haul truck and the job, price the mix, calculate the tax, print instructions to the driver, and accumulate the amount of mix on the project today and to date for the superintendent.

Fourth, at the end of the day, the computer can generate reports on production by mix, by company, by truck, by project, or by just about any field of data captured on the load ticket.

Computer automation speeds production, increases the plant throughput per day, and is a watchful eye over quality for the plant operator, the superintendent, and the end user of the mix. However, an equally accurate mix can still be made by pushing buttons and watching the scales. In fact, most automated systems are designed so that the plant can be run manually if necessary.

Although specifying agencies have encouraged the move to computer automation by requiring printed tickets and by requiring that out-of-tolerance lockouts be installed on the plant controls, many plants are still producing a quality product, one batch at a time, through the efforts of a qualified operator. Although it improves efficiency and guards against errors, computer automation is not required to make a quality hot mix.

Today's computerized batch plant control room (Figure 4.40) typically has the following plant control areas:

- A motor control console with start/stop switches for plant motors. These are often interlocked for safety and plant protection. For instance, some motors cannot be started unless previous conditions are met. Programmable logic controllers, which are computers designed to start and stop motors and sequence industrial automation, are often used in the motor control console circuits.
- A burner control console with start/stop switches, increase/decrease switches, and the safety circuits required for the burner.
- A cold feed control console with start/stop switches and increase/decrease switches to control the feed from the cold bins to the dryer. Many cold feed control consoles have manual/automatic selector switches which allow the operator to control the feed from each bin manually or to call a cold feed mix formula from memory, select the desired production rate, and automatically feed the dryer to match the batch cycle requirements. A separate computer is sometimes found dedicated to this type of cold feed automation.
- The computerized batch automation resides in a batch control console. A draft control and air pollution control console houses the exhaust fan, dryer draft, and cleaning controls for the air pollution control equipment.
- Mineral filler, baghouse fines return, and lime control panels can be included if that equipment is required on the plant.

►►Recycling With A Batch Plant

All recycling methods in batch plants utilize conductive heat transfer instead of convective heat transfer. Conductive heat transfer is accomplished by mixing cold aggregate with hot aggregate. Convective heat transfer is accomplished by exposing cold aggregate to hot gases.

In batch plants, regardless of the recycling method used, superheated hot virgin aggregate is used to heat the cold, moist reclaimed asphalt pavement (RAP). A brief discussion of some of the most popular batch plant recycling and heat transfer methods is described below.

Weigh Box Recycling Technique With the weigh box method of recycling, cold (unheated), moist RAP is added to the weigh hopper, where the batch controls weigh it as an additional aggregate material (Figure 4.41). The feed bin for the RAP and the elevated conveyor that is required to reach the weigh hopper typically have large motors and pneumatic clutches with brakes so they can be started and stopped instantly. This facilitates feeding just the right amount of RAP into the weigh box.

The RAP is then mixed with the superheated virgin aggregates in the weigh hopper at that point, and the conductive heat transfer occurs in the weigh hopper and the pugmill throughout the dry mix cycle. During the heat transfer process, a significant amount of steam is released from the RAP. The pugmill and weigh hopper area must be enclosed and vented to the emission control system to accommodate this large volume of steam—and the amount of dust that is invariably carried with it from the dry virgin aggregates.

While recycled mixes can theoretically be produced with up to 50% RAP content with this method, in day-to-day practical field conditions it is rare to see RAP percentages higher than 25% with the weigh box heat transfer method. This is because RAP moisture contents typically run in the 3-5% range, and elevating the aggregate temperature beyond 315°C (600°F) is difficult due to dryer limitations. Also, dryer exit gas temperatures can impose a practical limit if bag-

Figure 4.40 Automated Batch Plant Controls (Courtesy of GenTec/Stansteel)

houses are used on the plant. The HMA producer must be aware of the temperature limitations of the filter fabric used in the bags of the baghouse and control the aggregate temperature accordingly.

Pugmill Recycling Technique With Separate RAP Weigh Hopper This method of recycling utilizes a separate weigh hopper for the RAP, which empties its batch component into the pugmill. The method is gaining in popularity and has several advantages as listed below:

- By adding an additional weigh hopper to the batch facility, the RAP is conveyed into and weighed in its own hopper while the virgin asphalt and virgin aggregates are being weighed. This reduces the batching time slightly.
- The same heat transfer, steam release, and practical limits apply to this approach as apply to the weigh box method of batch plant recycling.
- During long production runs of recycled pavement, an increase in the production rate per hour can be achieved with the slightly shorter batch cycle time.
- There is less wear and tear on the equipment from abrupt starting and stopping.
- The weighing process can be done more carefully and accurately with a separate weigh hopper that is undisturbed by instant steam release.
- Typically, a high-speed slinger conveyor is used to convey RAP from the RAP weigh hopper to the pugmill, although a chute or high-speed screw conveyor can also be used.

Figure 4.41 Weigh Box Recycling Technique at a Batch Plant (Courtesy of GenTec)

Bucket Elevator Recycling Technique Another approach to batch plant recycling is gaining acceptance due to its simplicity, and because it eliminates the steam release typical of the mixer and pugmill heat transfer methods.

In the bucket elevator recycling method, cold, moist RAP is mixed with the superheated virgin aggregate as the aggregate exits the dryer and enters the bucket elevator.

The continuous steam release resulting from conductive heat transfer occurs in the buckets as the virgin aggregate/RAP mixture makes its way to the screen deck. The steam released from the RAP is carried away by the fugitive dust ductwork already fitted to the bucket elevator and screen deck.

Because the RAP is being continually blended with the virgin aggregate, belt scales are used on both the conveyor feeding virgin aggregate into the dryer and the conveyor feeding RAP into the bucket elevator. The scales insure the maintenance of a proper ratio of RAP to virgin aggregate.

Gradation control for mix production is accomplished in one of two ways. However, the process is different from that in a batching facility producing completely virgin mixes.

In the first method, the RAP and virgin aggregate are both screened together over the screen deck, and the composite mixture is separated into the different hot bins in the tower. Each hot bin is sampled for asphalt content and gradation. Extractions must be done to determine the asphalt content of the material in each hot bin. Gradations of the hot bin samples must then be evaluated and individual hot bin percentages calculated based on the recovered gradations from each supply bin. The asphalt content reclaimed from the RAP is then determined based on the extraction results, and the new liquid asphalt requirement for each batch is established. It must be assumed that the RAP is consistent not only in the recovered stone gradation, but in the asphalt content and particle size of the RAP itself.

Because this approach to mix production is more difficult than with weigh-box or pugmill injection methods, a second method, which utilizes a screen bypass is frequently applied. With this method, gradation is controlled at the cold feed bins feeding the dryer, as with drum mix style plants. The virgin aggregate/RAP mixture is stored in a single hot bin in the tower, then weighed up as one pre-blended mixture in the aggregate weigh hopper. The mixture is diverted into one bin, typically the Number 1 bin, using a bypass chute fitted between the elevator and the screen deck.

Many states allow this type of approach, but usually also require belt feeders with variable speed drives, speed displays, and total and proportional control over each feed bin. This is the same generic requirement used for feeder gradation control on drum mixer style plants.

Because the trip up the elevator is relatively short in time duration, and because the RAP must be dry before it passes over the screens (or is stored in the combined RAP/ aggregate bin), RAP percentages rarely run over 20% with this approach.

Introducing RAP into Heat Transfer Chamber or Dryer

This approach is essentially the same as the standard bucket elevator method, but the addition of a heat transfer chamber in the combustion area of the dryer allows RAP heat transfer to begin in the dryer shell. Another difference is that steam release also begins in the dryer shell. An advantage in this process is that the virgin aggregate and RAP are already combined as they exit the dryer and enter the elevator.

Higher percentages of RAP can be achieved (25%-35% is typical) than with standard bucket elevator methods, because the RAP has a longer period of time for heat transfer to be completed. All other aspects of the standard bucket elevator method apply to this recycling approach.

RAP Consistency

Regardless of whether one uses a batch or a drum mix plant, when producing recycled mixes using reclaimed asphalt pavement (RAP), one makes the assumption that after testing the RAP for gradation, asphalt content, and asphalt characteristics, these characteristics do not change. RAP is treated as one material source with consistent gradation, asphalt content, and asphalt characteristics. The new recycled HMA job mix formula is created taking these material characteristics into consideration.

This is an important point to keep in mind. RAP sources must be properly and randomly tested for consistency prior to producing the mix formula. If the consistency of the RAP is suspect, then the quality of the final hot mix will be suspect. If RAP consistency is variable, using a higher percentage will increase the probability that the final product will be out of specification.

No plant equipment or control equipment can change this inevitability. This is one reason why specifying agencies frequently ask for RAP sources to be stockpiled separately, and why RAP percentages are often limited in the final product.

One way to assure better consistency of the RAP is to split it into two sizes—a fine material and a coarse material. Usually the split is made on the 9.5 mm (3/8 inch) or the 4.75 mm (No. 4) screen. The material would then be proportioned into the mix as if it were two different sources of material.

Plant Guidelines

Regardless of whether a batch plant is manual, semiautomatic or completely automatic, certain basic plant components and functions must be observed regularly to ensure that the plant is capable of producing HMA that meets specifications and is indeed doing so. Below is a list of items that should be checked regularly at all types of batch plants:

1. Accurate proportioning of cold feed aggregates.
   - To ensure proper blend of materials to meet predetermined job mix formula.
   - To ensure the proper balance of material in the hot bins.
2. Scales zero properly and record accurately:
   - Scale lever systems kept clean.
   - All scale lever rods, knife edges, etc., shielded where possible.
   - Load cells of the proper size, properly aligned and hanging free.
3. Asphalt bucket tared properly and hanging free.
4. Aggregate weigh box tared properly and hanging free.
5. Mixer condition and function:
   - Mixer parts in good condition and adjusted to the proper mixing pattern.
   - Proper size batch being mixed.
6. Sufficient mixing time for dry and wet cycles.
7. Uniform asphalt and aggregate distribution in the pugmill.
8. Valve and gate leaks needing repair.
9. Proper aggregate and asphalt temperature when these materials are introduced into the weighing receptacles.
10. Worn or damaged screens.
11. Moisture content of aggregate after it leaves the dryer.
12. All proper safety requirements being met.

At plants where an automatic control panel is used, the following items should be added to the checklist:

1. Correct input or formula data.
2. Bin withdraw order in proper sequence.
3. Automatic switch in "on" position.
4. Mix timers correctly set.
5. All control switches in correct position.
6. Proper open time of pugmill discharge gates.

When a plant uses an automatic recording device, the following items should be checked regularly:

1. Printouts check accurately against material input and scales.
2. Aggregate printouts refer to proper bins.
3. Printout readings remain continuous.
4. File of printout readings kept for designated period of time.

Drum Mix Plants

➤➤ Introduction

This section describes functions unique to drum mix plants and presents information to show how a plant should operate in order to produce a paving mixture that meets job specifications. Upon completing this section of the manual, the reader should be able to:

- List the major components of a drum mix plant.
- Explain the purpose of each component.
- Describe how each component works.
- Outline the process as materials flow through a drum mix plant.
- Recognize potential problems that may occur and describe specific measures that should be taken to prevent such problems.
- Prescribe corrective measures to be taken in the event that deficiencies are detected in the mix.

Drum mixing is a relatively simple process of producing hot mix asphalt. The mixing drum, or drum mixer, which is the major component, and from which this type of plant gets its name, is very similar in appearance to a batch plant dryer drum. The difference between drum mix plants and batch plants is that in drum mix plants the aggregate is not only dried and heated within the drum, but also mixed with the asphalt binder. There are no gradation screens, hot bins, weigh hoppers, or pugmills in a drum mix plant. Aggregate gradation is controlled at the cold feed. Figure 4.42 shows typical drum mix plants – one stationary and one portable.

As the various aggregates are correctly proportioned at the cold feed and introduced into the drum for drying, the asphalt binder is also introduced into the drum. The rotation of the drum provides the mixing action that thoroughly blends the asphalt binder and the aggregates. As the mix is discharged from the drum, it is carried to a surge bin or storage silo and subsequently loaded into trucks.

➤➤ History Of Drum Mix Plants

Drum mixing of hot mix asphalt materials was originally introduced about 1910. More than one hundred small drum mix plants were operated until the mid-1930s, when they were replaced by continuous mix and batch-type plants of greater production capacity. The drum mixing process was resurrected in a revised form in the late 1960s.

In the ensuing years, drum mix plants, also called drum mixers and formerly dryer drum plants, became widely used in the hot mix asphalt industry. Introduced on a wide scale in the early 1970s, drum mix plants quickly gained popularity among contractors due to their portability, efficiency, and economy. Drum mixers also have the ability to produce large quantities of high quality mix at relatively low temperatures or at conventional temperatures.

The 1980s and 1990s saw changes in the configuration of the drum that significantly improved the efficiency of the heating and mixing process. There was also improvement in the methods of adding reclaimed asphalt pavement (RAP) to the mixture. In addition, improvement in the computer controlled weighing systems has enabled materials to be more efficiently controlled and cold feed systems to be checked and calibrated more quickly and accurately. These advances provide for more precision in the control of HMA production.

Figure 4.42 Drum Mix Plants – Portable Type (above – Courtesy of CMI)
And Stationary Type (below – Courtesy of Warren Paving)

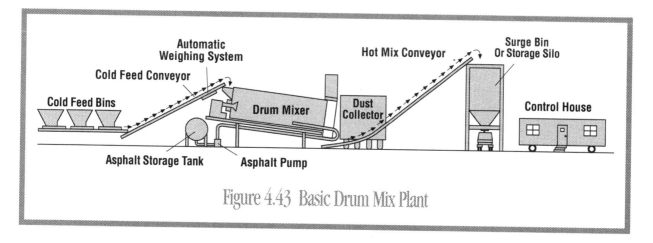

Figure 4.43 Basic Drum Mix Plant

➤➤ Drum Mix Plant Operations And Components

The fundamental components of the drum mix plant are:

- Aggregate cold feed bins
- Conveyor and aggregate weighing system
- Drum mixer
- Dust collection system
- Hot mix conveyor
- Mix surge bin or storage silo(s)
- Control house
- Asphalt Storage Tank(s)

Figure 4.43 shows the major components of a basic drum mix plant. Following is a brief, general description of the sequence of processes involved in a typical drum mix plant operation.

Controlled gradations of aggregates are deposited in the cold feed bins, from which they are fed individually in exact proportions onto a cold feed conveyor. An automatic aggregate weighing system monitors the amount of aggregate flowing into the drum mixer. The weighing system is interlocked with the controls on the asphalt storage pump, which draws asphalt from a storage tank and introduces it into the drum. The asphalt and aggregate are thoroughly blended by the drums rotating action. A dust collection system captures excess dust escaping from the drum. From the drum, the hot mix asphalt is transported by conveyor to a surge bin or storage silo, from which it is loaded into trucks and hauled to the paving site. All plant operations are monitored and controlled from instruments in the control house.

The mixing process is essentially the same in all parallel flow drum mix plants, which have been the standard type. Methods of feeding the materials to the drum, however, may differ. Other types of plants are described in later sections.

The production of HMA that meets job specifications is most easily accomplished when the various parts and functions of the plant are in balance; that is, when they are properly coordinated to work together as a smooth-working unit. Also essential for consistent high quality mix is uniform (uninterrupted) plant operation. Accurate proportioning of materials is entirely dependent on the uniform flow of those materials. Plant stops and starts adversely affect mix quality.

To ensure the balance and uniformity necessary to produce HMA that meets specifications in all respects, the following control equipment is required for all drum mix plants:

- Separate cold feed controls for each aggregate size and RAP.
- Separate indicators of material delivery rates for all components.

- Interlocking controls of aggregate cold feed, asphalt and additive delivery, and RAP delivery to the drum.
- Automatic burner controls.
- Dust collector that can feed back collected material into the system or waste the dust.
- Sensors to measure temperature of the hot mixture at drum discharge.
- Segregation controls on surge bins or storage silos.
- Moisture compensator.
- Alarms to alert the operator to any interruptions of material delivery.
- Equipment that will produce printout of material delivery rates and totals at specified intervals.

Controls and monitoring devices are usually located in the control house, where there is good visibility of the entire operation.

Aggregate Storage And Feed In a drum mix plant, mix gradation and uniformity are entirely dependent on the cold feed system. Proper care must be exercised not only in the production of the aggregate, but also in its storage. The contractor should provide for receiving and handling aggregates in such a way that there is no danger of contamination or intermingling. Among other things, this means providing clean surfaces on which to place the materials.

Stockpiles must have the proper gradation to meet final mix specifications, and they should be free of segregation. Enough stockpiles of different-size fractions should be maintained to properly control the gradation of the mix. Practices vary with respect to the sizes of aggregates that are separated into different stockpiles. Without this stockpile separation, it may be difficult to maintain the proper gradation control.

Segregation can be prevented by constructing stockpiles in layers not exceeding 1.2 meters (4 feet), and loading aggregate from the upper areas of the stockpile, thereby minimizing sloughing of the side slopes.

Segregated stockpiles, if uncorrected prior to entrance into the mixing plant, will result in mix gradation difficulties. The plant operator should establish and maintain non-segregated stockpiles. If the stockpiles are segregated, through poor stockpiling practices or negligence, the deficiencies in uniformity must be corrected before the aggregate is fed into the mixing plant. Regardless of the method of handling, all efforts should be directed at delivering the correct, uniformly graded aggregate blend to the mixing plant.

Since the typical drum mix plant, unlike a batch plant, does not incorporate a gradation screening unit, the aggregate must be correctly proportioned prior to its entry into the mixing drum. This is essential. The most efficient way to accomplish this is with a multiple-bin cold feed system equipped with precision belt feeders for the control of each aggregate. Under each bin is a belt feeder onto which the aggregate is dropped. Precise controls (Figure 4.44) are used here to feed the exact proportions from the belt feeders onto the belt.

The plant should be equipped with provisions to conveniently obtain representative samples of the full flow of material from each cold feed and the total cold feed. Routine samples are obtained from these points and tested to determine gradation.

Cold feed control consists of the following:

1. Performing a sieve analysis of the aggregate in each bin
2. Calibration of the feeders – both gate opening and belt speed
3. Establishment of bin proportions
4. Setting the gate openings and belt drive speeds

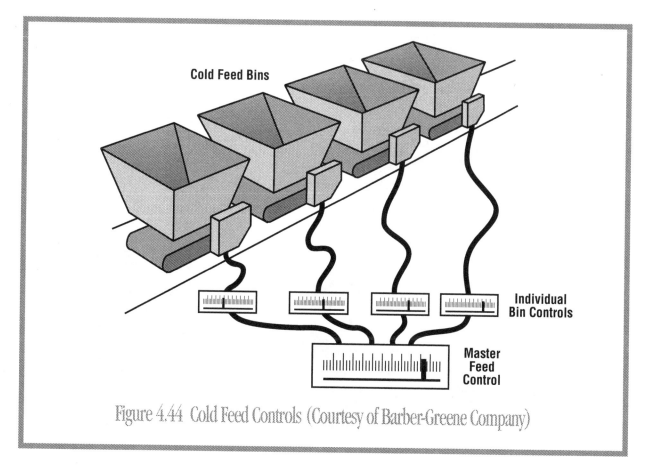

Cold Feed Bins

Individual Bin Controls

Master Feed Control

Figure 4.44 Cold Feed Controls (Courtesy of Barber-Greene Company)

Once calibrated, the gate openings should be checked frequently to ensure that they remain properly set. All settings should be considered tentative because the cold aggregate used in the mix may vary in grading and moisture content. This may require adjustment of the gates to maintain a uniform gradation.

To calibrate the aggregate metering system and to plot a cold feed capacity chart, a sampling device or method to obtain samples from the belt is necessary. The device should permit the flow of aggregate to be diverted to a collection container for accurate weight checks of timed aggregate samples (Figure 4.45). Such devices are usually installed at the end of the conveyor belt just prior to entry into the drum mixer.

Drum mix plants require a continuous weighing system on the cold feed conveyor belt. In-line belt weighers, also called weigh bridges (Figure 4.46), are continuous belt-weighing devices used for this purpose. Combined aggregates passing over the conveyor belt are continuously weighed, and a readout in the control house indicates the weight of the flow over the scale at any given instant. No material should be diverted from the conveyor belt, other than samples, after it passes the belt weigher.

The in-line belt weigher is usually located midway between the head and tail pulley of the cold feed belt conveyor. This location tends to lessen variations in reading caused by impact loading, rollback of aggregate, or changes in belt tension. Means can be provided for conveniently diverting aggregates into trucks, front-end loaders, or other containers for checking the accuracy of the belt weigher. The device should be accurate within + 0.5 percent when tested for accuracy.

In drum mix plants, the aggregate is weighed on the belt before drying. Since the undried material may contain an appreciable amount of moisture that can influence the aggregate's

Figure 4.45 Typical Sampling Device (Courtesy of Astec)

Figure 4.46 In-Line Belt Weigher (Weigh Bridge) (Courtesy of Gencor)

weight, an accurate measurement of aggregate moisture content is important. From that measurement, adjustments can be made to the automatic asphalt metering system to ensure that the amount of asphalt delivered to the drum is proper for the amount of aggregate minus its moisture content.

The moisture content of the cold feed aggregate should be checked before beginning each day's operation and again about the middle of the day. The contractor should adjust the moisture control equipment accordingly. If the moisture content is believed to vary during the day, it should be checked more frequently. The moisture content may be determined manually or electronically. Provisions should be made for electronically correcting wet aggregate weight readings to dry aggregate weight readings continuously during production.

Modern plants have moisture-sensing devices that can sense changes in moisture content once set to a specific moisture content. The device is coupled into the control system and compensates for moisture changes automatically.

Asphalt Control With A Drum Mixer Asphalt control in a drum mixer operation is done continuously. The asphalt is measured through a calibrated meter on a continuous basis, relative to aggregate flow, and then combined with the aggregate in the drum.

The asphalt content is interlocked to the aggregate flow. The monitoring system notes changes in the weight of aggregate over the belt scale and automatically adjusts the asphalt flow to compensate for these changes. The belt scale measures the aggregate flow and reports to the control circuits. The asphalt meter measures the asphalt flow and reports to the control circuits. The control circuits compare the two numbers against the set point established by the operator. If an adjustment is required, then a signal is sent to the asphalt flow control mechanism to speed up or slow down the asphalt flow.

➤➤ Drum Mix Operation

Basic Description Drum mix plants make use of a process whereby aggregates in their proper proportions are fed to a dryer drum, dried and heated in that drum, and then mixed with asphalt binder and continuously fed to a surge bin or storage silo. There are a number of ways in which this process can take place, and these will be described. However, there are some basic, common operations that will first be discussed.

The heart of the drum mix plant is the drum dryer/mixer itself. It is similar in design and construction to a conventional batch plant rotary dryer, except that a drum mixer not only dries the aggregate but also blends the aggregate and asphalt together into a hot mixture.

Aggregate enters the primary zone of the drum, where heat from the burner dries and heats it. The aggregate then moves to a secondary zone, either in the same drum or a related location, where asphalt binder is added and the two are thoroughly blended. The mixture of hot asphalt and the moisture released from the aggregate produces a foaming mass that traps much of the fine material (dust) and coats the larger particles.

It is important that the aggregate in the drum not only rotate with the revolving motion of the drum, but also spread out sufficiently to make heating and drying of all particles quick and efficient. Drum mixers are equipped with specially-designed flights to create a "veil" of aggregate at appropriate parts of the drum for obtaining maximum drying and minimum exhaust gases.

Asphalt binder is pumped from the asphalt storage tank and enters the mixing drum at the appropriate point in accordance with the design of the process. When the asphalt binder is added into the drum, it is pumped into the bottom of the drum at about the same location that

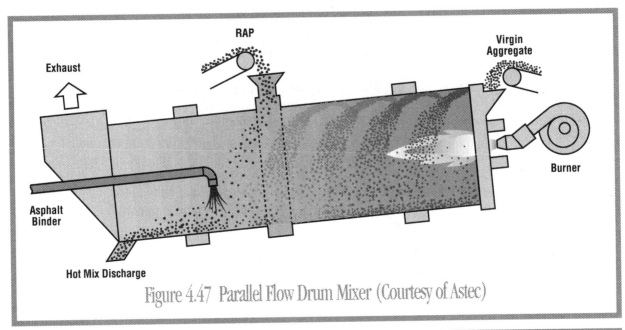

Figure 4.47 Parallel Flow Drum Mixer (Courtesy of Astec)

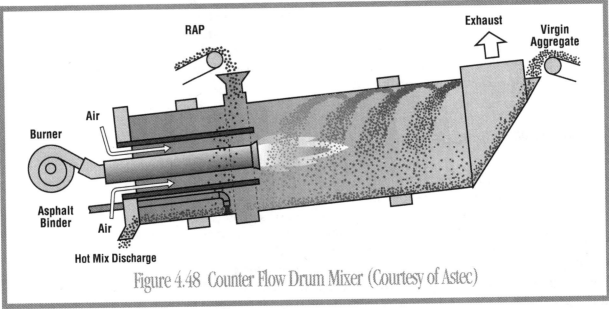

Figure 4.48 Counter Flow Drum Mixer (Courtesy of Astec)

the mineral filler and/or baghouse fines are introduced. Adding asphalt binder and dust in close proximity allows the asphalt to trap a good portion of the fines and coat them before they are picked up by the high-velocity exhaust gas stream. The exhaust gases are passed through a dust collection system where enough of the remaining dust is trapped and removed to meet emission requirements.

Mix temperature is monitored continuously by a sensing device at the discharge end of the drum mixer. The temperature recorder and other indicators are located in the control house. A suitable means can be provided for inspecting and sampling the HMA at the discharge end of the drum.

Burner Operation and Control The purpose of the burner inside the drum mixer is to provide the heat necessary to heat and dry the aggregate used in the final mixture. The burner provides this heat by burning fuel – normally oil, natural gas, propane, or coal. The controls for the burner are located in the control house.

When oil is burned, low-pressure air drafts are used to atomize the fuel oil for burning. Burners using natural gas and propane can be low-pressure or high-pressure units. In all cases, the fuel feed and air blower must be balanced to ensure that the proper proportions of fuel and air are being introduced into the burner for efficient combustion. Lack of balance can lead to incomplete burning of the fuel, which, especially in the case of fuel oil or diesel fuel, can leave an oily coating on the aggregate particles. Such imbalances between fuel feed and airflow can be corrected by either decreasing the fuel feed rate or increasing the blower or draft air.

Bins, Silos and Weigh Systems In a drum mix operation, because hot mix asphalt is produced in a continuous flow, it is necessary to have a surge bin or storage silo for temporary storage of the material and for controlled loading of trucks. The weight of HMA in the trucks is normally determined by the use of platform scales. The weigh system control panel located in the control house typically records weight measurements.

►►*Different Designs of Drum Mix Plants*

Parallel Flow The original drum mix plants were of the parallel flow type, and for many years they were the mainstay of drum mixers. In the parallel flow plant, aggregate enters the drum on the burner side and flows in the same direction as (parallel to) the hot gases. Asphalt binder is introduced about one-third of the drum length from the discharge end, as are any additional fines (dust) necessary for the mix.

The percentage of parallel flow plants has decreased significantly in recent years, even though many are still in existence and some new ones are being produced. The parallel flow process is shown in Figure 4.47.

Counter Flow In the counter flow drum mix plant, aggregate enters the drum through the end opposite the burner and flows in the opposite direction of the exhaust gases. The burner is extended an appropriate distance into the drum so that the burner flame is introduced beyond the point where the binder is injected and mixing of aggregate and asphalt take place. Figure 4.48 illustrates the counter flow drum mixing process with the mixing performed 0in the same drum where the drying is accomplished. The counter flow drying process is also used in other variations of drum mix plants.

Coater Units A "coater" unit can be added to the discharge end of a parallel flow or counter flow drum mixer. The aggregate is heated and dried in the drum and transferred to the coater for mixing with asphalt binder. The asphalt can be injected at the extreme end of the dryer or into the coater. The advantage of the coater is that it takes the mixing process out of the path of the hot exhaust gases. It also avoids potential steam distillation of the asphalt binder. The coater can be either a single or double shaft pugmill or an auger mixer. Figure 4.49A shows a parallel flow drum mixer with a coater and Figure 4.49B shows a counter flow drum mixer with a coater.

Double Barrel The Double Barrel® drum mix plant utilizes a counter flow dryer as an "inner" drum. A larger diameter "outer" drum covers about two-thirds of the inner drum and forms a mixing chamber, where the aggregate and asphalt binder are mixed after the aggregate is dried

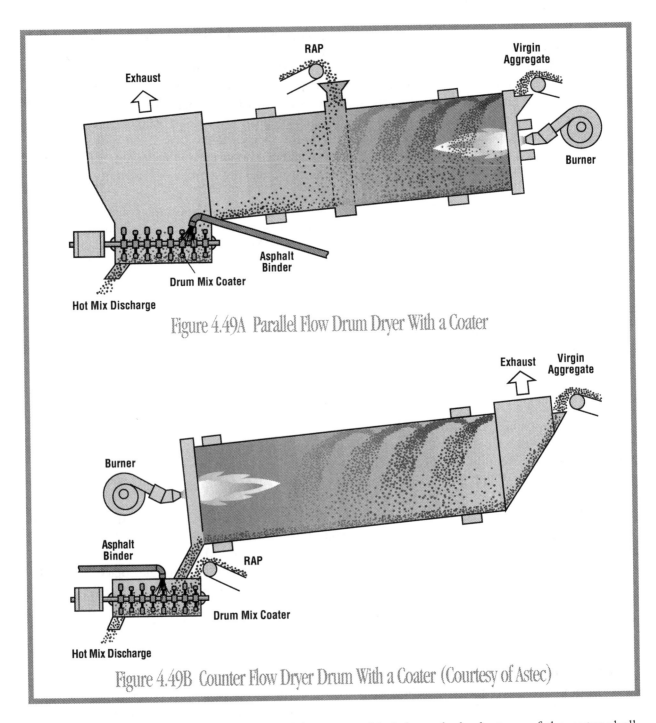

Figure 4.49A Parallel Flow Drum Dryer With a Coater

Figure 4.49B Counter Flow Dryer Drum With a Coater (Courtesy of Astec)

in the inner drum. The asphalt binder and fines are added through the bottom of the outer shell into the mixing chamber, and away from the flow of hot gases in the dryer. Figure 4.50 shows this configuration.

Triple Drum The Triple-Drum™ plant is a drum mix plant that features an extended counter flow drum that contains three distinct zones for drying and heating and for mixing. The aggregate is dried and heated in the first and longest zone. In the second zone (where the burner is located) the aggregate flows partly in the inner drum (under the flame) and partly in the chamber formed by the outer drum around the flame. It gets additional heating as it moves past the

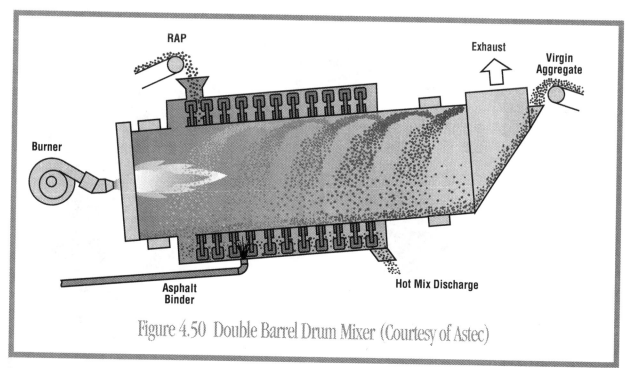

Figure 4.50 Double Barrel Drum Mixer (Courtesy of Astec)

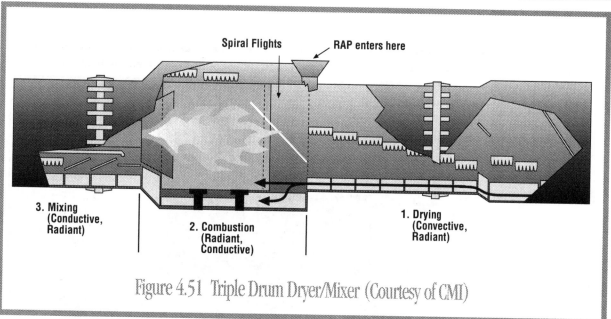

Figure 4.51 Triple Drum Dryer/Mixer (Courtesy of CMI)

flame area. When the aggregate reaches the third zone, asphalt binder is added and mixing takes place. This process is shown in Figure 4.51.

Double Drum The Double Drum™ plant makes use of two separate drums, a counter flow drum for drying and heating the aggregate and a smaller rotary mixer for combining the aggregate and asphalt binder. The aggregate flows from the dryer drum to the rotary mixer, where the asphalt and fines are injected near the entrance. This type of drum mix plant is shown in Figure 4.52.

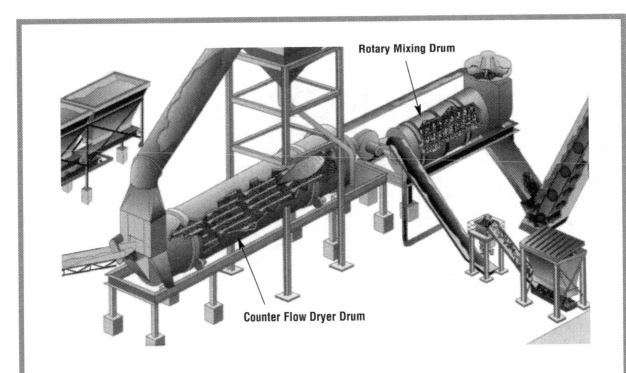

Rotary Mixing Drum

Counter Flow Dryer Drum

Figure 4.52 Double Drum Dryer/Mixer (Courtesy of GenTec)

Figure 4.53 RAP Introduced Into a Drum Mix Plant (Courtesy of Astec)

➤➤Recycling With A Drum Mix Plant

The introduction of reclaimed asphalt pavement (RAP) into a drum mixing operation is quite different from the methods used in a batch plant. RAP is added in its natural state into the drum mix process so that it can receive heat and combine with the superheated aggregate prior to the addition of asphalt binder. This softens the old asphalt and heats the RAP. At the same time the point of entry of the RAP is kept as far away from the flame as practical. Steam is released when the RAP comes in contact with the superheated aggregate. The quantity and moisture content of the RAP determines the intensity of the steam. The quality of the resulting recycled mix depends on the quality of all the materials used (including RAP), the proportions used, and the method in which they are combined. Figure 4.53 shows RAP being introduced into a drum mix plant.

In the parallel flow drum mixer, RAP is added through a collar in the top of the drum located just beyond its midpoint (Figure 4.47). RAP is added to the counter flow drum mixer through a collar at a point just behind the flame, and approximately two-thirds of the distance downstream from where the aggregate enters (Figure 4.48).

The parallel flow dryer with a coater receives the RAP in the same manner, and location as the parallel flow drum mixer (Figure 4.49A). The counter flow dryer with a coater receives the RAP at the entrance of the coater (Figure 4.49B).

The Double Barrel drum mixer uses an entrance port in the top of the outer drum on the end nearest the flame to add the RAP (Figure 4.50). This introduces the RAP into the mixing chamber before the asphalt binder, allowing the hot aggregate and the heat from the inner drum to heat the RAP.

In the Triple-Drum plant, RAP enters the drum near its midpoint and through an opening in the top of the outer drum (Figure 4.51). RAP is mixed with virgin aggregate as it is carried past the flame in a chamber formed by the inner drum, which surrounds the flame,z and the outer drum.

The rotary mixer, or second drum, is where RAP is added in the double drum plant (Figure 4.52). The RAP enters at the end of the mixer at about the same point that the heated aggregate and asphalt binder are introduced.

CHAPTER 5

Objectives At the conclusion of this chapter, one should be familiar with basic asphalt paving operations under normal paving conditions and:

- Understand all procedures for placing hot mix asphalt (HMA).
- Be familiar with the requirements of surface preparation prior to paving.
- Understand the principles of the asphalt paver and the floating screed.
- Understand the principles and functions of an automatic screed control.
- Understand how to set proper grade to meet smoothness specifications.
- Know how to plan and control a paving operation.
- Understand placing and mix deficiencies and how they might be corrected.
- Know how to match and/or construct transverse and longitudinal joints.

Introduction

➤➤ General

The knowledge of materials, construction and equipment are highly important in the handling and placement of HMA. The relationship of materials properties and their behavior in the placement operation must be thoroughly understood to obtain a durable, smooth-riding pavement that meets all the standards of quality construction.

To achieve a quality pavement, a cordial and cooperative relationship between the agency and the contractor is imperative. Each party must understand the project specifications and the scope of the work. Agency and contractor must be familiar with the equipment necessary to perform paving operations and the proper use of the paving equipment.

Because communication is essential for successful paving operations, a pre-construction conference should be held prior to each paving project. Such a conference allows the project engineer, the paving superintendent, the contractor, the agency representative and others directly involved with the operation an opportunity to discuss topics such as:

- Project chain of command.
- Project schedule, specifications & testing requirements.
- Alternative procedures to the plans.
- Quality control plans; reasons for rejection/acceptance of work.
- Use of new equipment or test methods.
- Plant and paver production rates.
- Traffic control.
- Record keeping.
- Special equipment or personnel requirements.
- Method for determining density compliance.
- Paving plan: production rate, haul distance, number of trucks, continuity of operation.

The pre-construction conference is the time for questions to be answered, potential problems to be resolved and channels of communication and command to be established. It is a time to establish relationships with everyone involved in the project so that confusion and friction can be avoided once paving operations begin.

➤➤Paving Crew

A typical paving crew is a team that includes individuals having special interest in the operation. Typical members of this team are: paving superintendent, paver operator, dump person, screed person, and a couple of people to lute and take care of joints and mat repairs. The superintendent looks after the over-all operation while the other crew members are charged with specific details. The paver operator is concerned with the paver and its operation. The "dump" person positions himself in front of the paver so contact is maintained with the paver operator and the truck drivers. The dump person guides the paver into the pickup position of the truck and assists in dumping the material into the paver hopper. This person may also receive and review the load tickets for load time, temperature and tonnage. The screed person operates the screed setting for proper depth, smoothness, and texture, while confining the mix so compaction will be obtainable. Coordination of the entire crew with the paver operator is essential to achieve all the desired goals.

➤➤Project Management Responsibilities

The project manager should be thoroughly familiar with the specifications and see that they are complied with during the paving operation. His responsibilities include the following:

- Ensure that each load of mix is satisfactory, that data from the truck tickets is recorded accurately, and that the paver is operated properly. Should any deficiencies appear in the mat during placing, corrections are to be made before the mix cools.
- See that attention is given to details such as proper thickness of the spread, proper section, joints, surface texture and uniformity.
- Monitor the temperature of the HMA to ensure that proper mix and compaction temperatures are maintained during the paving operation.
- Be courteous and tactful in dealings with the contractor and in requests for corrective action when necessary.
- Assure that accurate, detailed records and diaries are kept. In addition to the information included on haul tickets, there are other important items that must be recorded as part of the permanent records. Any unusual occurrences or changes in construction methods, equipment, appearance, or handling properties of the mix should be noted in the diary along with the station (location) on the roadway where the change was made. The daily diary is for the project manager's information during construction, but upon completion of construction it should become a part of the permanent records of the project. A typical page from an inspector's diary is shown in Figure 5.01.

In addition to the information shown above, the project manager should assure that the following are included in the project records for each type of mixture used: test reports on mixture analysis, reports on pavement thickness, smoothness and density of the finished pavement. The inspector's records should also note any delays in the work and their causes, as well as a list of all visitors to the project.

```
CMCAR032                          ODOT CONSTRUCTION MANAGEMENT SYSTEM                    12/22/97  14:05
                                     DAILY DIARY FOR 09/11/97                                 PAGE:     1

Project: 97-0421  County: PER  Route: SR-668   Section: 000.00  PE/PS: WILLIAM R2  District: 05  Diary Entry Date: 09/12/97
Diary Approved: Y  Approval Date: 09/22/97  Potential Claim: N

GENERAL INFORMATION
Temp: 60/70                       Precip: PM SHOWERS              Lost Day Due to Weather: N  Lost Day Due to Other Reasons: N

General Remarks: CONTRACTOR PLACING SURFACE COURSE 448 TYPE 1 TODAY.         CONTRACTOR FOLLOWED UP WITH TEMP CENTER LINE.
                 CONTRACTOR WAS RAINED OUT AT 2:00 PM TODAY

CONTRACTOR
Name: SHELLY COMPANY THE (INC)              Superintendent: DICK BORING      Work Hours: 7:00/7:00
Description of Work: 448 TYPE 1 SURFACE COURSE
Worker Breakdown: Supervisors  1  Skilled  4  Other  4  Alt1:  0  Alt2:  0  Alt3:  0  Total Personnel for the Day      9
Equipment Breakdown:  Equipment Number  Equipment Type                                    Idle
                                        PICKUP 2                                           N
                                        RUBBER ROLLER                                      N
                                        BROOM                                              N
                                        BERM BOX                                           Y
                                        TACK TRUCK                                         N
                                        WATER TRUCK                                        N
                                        STEEL ROLLER                                       N
                                        THREE WHEEL ROLLER                                 N
                                        PAVER                                              N

STATE EMPLOYEES HOURS WORKED
Employee Name                       Hours Worked   Work Code   License Number
THOMAS            BOOKMAN               8.0           5192        T5-874
TERRY            E WOLLENBERG           0.0           5192        T5-696

REFERENCE NUMBERS FOR PAY
Ref Num  EW Num   Pt Cd   Item Code   Item Desc                    Location              Quantity/Amt ($)   Unit   ID
  17               01     448E16000   ASPH CON SURF TP 1           144+25-274+50 RT LFT       557.050        CY    TB
  33               01     614E21400   TEMP CL, CL II               144+25-236+75               1.750        MILE  TB
```

Figure 5.01 Typical Page From an Inspector's Diary

►►Surface Inspection and Preparation

HMA can be placed on a variety of surfaces, including:

- Subgrade (soil)
- Granular base course (aggregate)
- Existing asphalt pavement
- Existing portland cement concrete pavement
- Rubblized pcc pavement
- Brick pavement

Certain inspection and control procedures are common to the preparation of all of these surfaces. Others pertain only to one or two types of surfaces. Items to be performed and checked for each type are described below.

Subgrade The subgrade (soil) under a pavement is the foundation for the pavement. Regardless of the type of pavement to be placed, the subgrade must meet certain specification requirements. It must be properly graded to line and grade according to the plans. It must be thoroughly and uniformly compacted to the required density.

PLACING HOT MIX ASPHALT   5-3

The subgrade should be inspected to identify areas of soft or yielding soil that are too weak to properly support the paving equipment and haul trucks. Such areas should be corrected prior to paving. Checks should be made of both the transverse and longitudinal grade. If either is not within tolerance, the grade must be corrected, by removing material, or by adding and compacting material similar to that already in place.

If a Full-Depth® asphalt pavement is to be placed, special care must be taken in preparing the subgrade. A Full-Depth asphalt pavement is one in which asphalt mixtures are used for all courses (layers) above the subgrade. The HMA base is placed directly on the prepared subgrade. For a Full-Depth asphalt pavement to be placed properly, the subgrade surface must be firm, hard and unyielding to achieve proper density in the HMA course. It should be free of loose particles and accumulations of dust. If the subgrade has dried and becomes loose and dusty due to construction traffic, it should be watered, lightly bladed and rolled prior to placement of the HMA course.

Base Course A base course can be either a layer of granular material (aggregate) placed on the subgrade and compacted or, in the case of Full-Depth asphalt pavement, a layer of asphalt concrete. In either case, the base must be uniform in strength and within grade tolerances as required by the specifications. In addition, the surface should be free of debris and accumulations of dust. If the base course is an unbound material and has been primed with a cutback asphalt, it should be cured and free of diluent. The base is then swept lightly with a power broom to remove loose particles from the surface prior to placement of the HMA course. If the base course is an HMA base it should be swept with a power broom for removal of loose particles and dust. A tack coat is applied just prior to placing subsequent pavement layers.

Asphalt Pavement When HMA is placed over an existing asphalt pavement, or any other pavement for that matter, it is called an HMA overlay, or an asphalt overlay. An HMA overlay is designed to rehabilitate and strengthen an old pavement, extending its life and correcting surface irregularities.

When an existing asphalt pavement is covered with an HMA overlay, it must be properly prepared to ensure satisfactory performance of the overlay. Potholes and unstable sections must be repaired. Slight depressions must be cleaned out and filled with new material; deep depressions should be cut out of the pavement and replaced with new material. If the base course or the subgrade beneath a section of the old pavement is unsuitable, it too should be removed and replaced. Cracks greater than 3mm (1/8-in.) must be cleaned and filled prior to overlaying. Care should be exercised to not overfill the cracks. Overfilling can result in slippage of the material or result in bleeding of the sealant through the new overlay. When thin surfacing courses are to be used, it is especially important that prior correction of the surface contour be made.

When overlaying a rutted pavement, the cause of rutting must be determined. If the rutting is caused by:

- A plastic mixture(s), that material should be milled off prior to the placement of new HMA
- Consolidation of base layers or subgrade, it can be milled to the depth of the rut and overlaid. Another option would be to place a scratch or leveling course in the rutted areas, using a pneumatic tire roller as the intermediate roller in compaction.
- Structural failure of a pavement section, the section should be redesigned based on traffic (EALs), subgrade strength, and environmental factors.

Figure 5.02 Milling Machine (Courtesy of ROADTEC)

Occasional high spots can be removed by cold milling to a predetermined depth. Cold milling machines can be operated from a fixed grade-line, or a traveling grade-line can be attached to the machine to re-establish the profile grade and pavement section (Figure 5.02).

When only minor surface distortion exists, the construction of spot leveling courses can be used to restore proper line and cross-section. Milling may also be required in areas where minimum clearance or the matching of an existing elevation is to be maintained. All milled areas should be thoroughly cleaned with a power broom and tacked at the upper rate of the recommended application rate.

Leveling wedges of HMA are used to level depressions in an old pavement prior to the surfacing operation. The placing of leveling wedges is part of the leveling course operation. Leveling wedges should be placed in two layers if they are from 75 to 150 mm (3 to 6 inches) in thickness. Wedges thicker than 150mm (6 in.) should be placed in compacted layers of not more than 75 mm (3 in.). In placing multiple layers the shortest length layer should be placed first, with the successive layer or layers extending over or covering the short ones. (See Figure 5.03 for illustrations of correct and incorrect ways of making leveling wedges.) If the incorrect method is used, as shown in the lower illustration, there will be a tendency for a series of steps to develop at each joint because of the difficulty of feathering out asphalt mixtures at the beginning and end of a layer. A bump at these joints will reflect through to the final surface.

Where wedging of dips requires multiple layers, a survey should be made to plot profiles and cross-sections accurately. From these, the grade of the proposed correction and the lineal limits of the successive layers should be determined so that the project personnel can be given definite stationing for starting and terminating the spreader or motor grader passes (See Figure 5.04). Figure 5.05 illustrates the correct way to place leveling wedges for overcoming excessive crown.

Correct

Incorrect

Figure 5.03 Correctly Placed Leveling Wedges Ensure Smoother Pavements

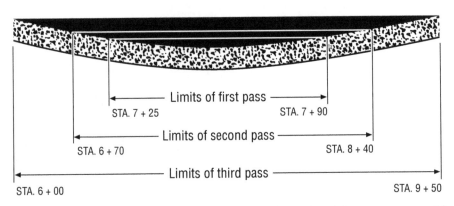

Limits of first pass
STA. 7 + 25   STA. 7 + 90
Limits of second pass
STA. 6 + 70   STA. 8 + 40
Limits of third pass
STA. 6 + 00   STA. 9 + 50

Figure 5.04 Limits for Multiple-Layer Leveling Wedges Should Be Determined by Level

Figure 5.05 Correctly Placed Leveling Wedges for Overcoming Excessive Crown

Caution No asphalt materials should be placed on an existing pavement containing excessive moisture. All existing surfaces should be dry and free of moisture. Moisture trapped deep in the pavement layer could cause moisture damage in the underlying pavement layers, resulting in the loss of stability and rutting. Moisture from the emulsified tack material is not considered a problem as this is surface moisture and not internal.

Portland Cement Concrete (pcc) Pavement Overlaying portland cement concrete pavement demands special preparation. First, a determination must be made as to whether the existing pavement slabs are in a suitable state of repair, and if they are stable or moving under traffic as a result of loss of subgrade support. All joints which are uneven or in need of repair and all unstable slabs must be identified. Slabs must be repaired or stabilized either by undersealing or by one of the fractured slab techniques, i.e. cracking and seating, breaking and seating or rubblization. Pavement evaluation by the Falling Weight Deflectometer (FWD) can be used to determine which option is best suited for the situation. Fractured slab techniques are the only options that will substantially prevent inherent pcc distress from reflecting through the asphalt overly.

Undersealing Undersealing is a method for stabilizing moving slabs and filling voids that may have developed under them. This process is normally bid by the gallons of asphalt actually used in stabilizing the slabs. Generally, this operation includes:

- Drilling holes in the depressed or unstable slab.
- Filling voids under the slab with specified asphalt cement for undersealing. The asphalt is heated and pumped under the slab by pressure pumps.
- Plugging holes in the slab with wooden plugs until the asphalt underneath cools and solidifies.
- Filling holes with grout mix.
- Removing any excess asphalt material from the slab.

The same procedure can also be performed using portland cement grout or other suitable material.

Cracking and Seating; Breaking and Seating Cracking and seating and breaking and seating are essentially the same process, with the only difference being the type of pavement on which the process is applied. Cracking is the term used when the pavement being treated contains no steel, such as plain jointed concrete pavement. Breaking is the term applied to pavements that are reinforced, either with mesh or reinforcing steel. This process is more severe in that the concrete bond must be broken from the steel. In either case the unstable slab is fractured into smaller pieces and then firmly seated into the underlying base. Any existing HMA overlay must first be removed before the cracking or breaking and seating procedure can be used. The process is normally bid by the square meter (square yard) of pcc cracked or broken and seated. For more information, refer to the *Asphalt Institute technical bulletin, Cracking and Seating Prior to Overlay with Asphalt Concrete* (TB-1).

Generally, the procedure is as follows:

- A drainage system consisting of edge drains or underdrains parallel to the pavement centerline is installed.
- The slab is cracked or broken into smaller pieces with a drop hammer or other pavement breaking machine.
- The pieces are seated into the underlying base by rolling them with a heavy roller of specified weight.
- The seated pieces are tacked, if required, and covered with HMA.

Rubblization Rubblization (or rubblizing) is the process of fracturing the slab into smaller pieces, generally 25 to 150 mm (1 to 6 inches) in size, and effectively reducing the pcc pavement to a crushed aggregate base. The installation of a drainage system, similar to that described in the cracking/breaking procedure, is required. The rubblization process is accomplished by the use of a high frequency, low amplitude resonant breaker or impact breaker. Compaction is typically accomplished with a vibratory roller and/or a pneumatic-tired roller. Traffic should be prohibited from using the prepared surface after the compaction has been completed. The application of a tack coat is not required if the paving is to be done immediately after the rubblization because the HMA delivery trucks will pick it up on their tires and damage the finished surface. A 37-mm to 50-mm (1.5-in. to 2-in.) leveling course on the rubblized pavement is often helpful in obtaining a smooth finished pavement.

When undersealing, or fractured slab procedures are used, accurate records must be kept of the number of gallons of asphalt used for underseal or the number of square meters (square yards) of pavement cracked, broken or rubblized. The volume of asphalt used in the undersealing is calculated at 15°C (60°F). The records are necessary for determining how much the contractor is to be paid for these items.

▶▶ Adjusting Street Fixtures

Before initiating an overlay operation, manholes, catch basins and utility appurtenances must be raised so that they will be level with the surface. In the case of manholes, ring-type collars are used to raise the height. For catch basins, extensions may be shop built to the proper height. If grade changes are extensive the appurtenances may require reconstruction. After they are raised, but prior to the overlay being placed, exposed fixtures must be marked by flags or barricades so as not to present a hazard to traffic.

▶▶ Horizontal and Vertical Grade Control

In new pavement construction both horizontal and vertical grade control must be established to ensure that the finished pavement is constructed in accordance with the project plan location and profile. A survey crew will normally establish the profile grade line (PGL) of the proposed pavement and set grade stakes (blue tops) in the subgrade parallel to, and at a fixed distance from the PGL on either side of the roadway. On tangent sections of road, the stakes are usually placed at 30-meter (100-ft.) intervals. They are set closer on curves. In setting the initial grade stakes, consideration of the over-all deviation from the section must be considered. Typical grade stakes are set to allow the contractor no subgrade too high and none more than 12.5 mm (1/2 in.) low from the PGL on the subgrade. Each subsequent pavement layer will have a different tolerance for the layer thickness.

Although it is the responsibility of the contractor to place stakes for line and grade, it is the duty of the agency representative to check the accuracy of such.

Transverse grades can be checked by stretching a stringline of known height above and between the control stakes (blue tops). The distance from the stringline to the grade should be uniform (Figure 5.06). Differences should be noted as cut or fill (high or low).

Either building up or cutting away sections of subgrade constitutes grade correction. The motor grader equipped with a laser control is the normal piece of equipment used in obtaining the true grade-line. However, in small areas, hand work may be required to trim around appurtenances. Where irregular roadway surfaces require a high degree of leveling, the best method to accomplish this is by using electronic grade controls on the grading equipment or the paver.

In establishing such a reference line, a wire or string capable of at least 125 pounds tension is erected on grade support standards. The line is stretched and anchored at intervals of 90 to 150 m (300 to 500 feet) and supported at 8m (25-ft.) centers. If there are sharp changes in

Figure 5.06 Checking Subgrade Elevations

grade, or on vertical curves, it is necessary to shorten the distance between the line anchors. As temperatures change throughout the day, the tension must be checked as increasing temperatures may cause sag. It is also necessary to locate the line in an area where traffic will not interfere with it. On curves, the distance between anchors should be shortened and the supports placed at closer intervals to maintain the alignment of the curve.

►►Prime Coats and Tack Coats

Prime coats and tack coats are applications of liquid asphalt applied to base material or lower layers of the pavement.

Prime Coats A prime coat is a sprayed application of a medium curing cutback asphalt or emulsified asphalt prime (EAP) applied to a base course of untreated material. When a medium-curing cutback asphalt is used, it is applied heavily enough to penetrate into the base material. An EAP penetrates more slowly and is applied at a lesser rate. A mixing grade asphalt emulsion can also be used, but it must be mixed into the base material by a motor grader, or rotary mixer type equipment. This can be done at the time of final grading and rolling of the base material. Prime applications are done under the same general weather conditions as paving; however, for cutbacks to properly cure, 24 to 72 hours of favorable weather are required.

If the HMA pavement is less than 100 mm (4 in.) thick, a prime coat is recommended prior to placement of the mixture unless prevailing circumstances prohibit it. Circumstances of this nature are when foot traffic is present, or there is a strong possibility of run-off, or the project cannot be closed for proper curing time.

A prime coat has three purposes:

- It fills the surface voids and protects the base from weather.
- It stabilizes the fines and preserves the base material.
- It promotes bonding to the subsequent pavement layers.

Application rates for prime coats vary with the type of asphalt used. For a medium-curing cutback asphalt, MC-30, 70 or 250, the application rate ranges from 0.9 to 2.3 liters/m² (0.2 to 0.5 gals/yd²). When an emulsified asphalt SS-1, SS-lh, CSS-1 or CSS-lh is used, application rates vary from 0.5 to 1.4 liters/m² per 25mm of depth (0.1 to 0.3 gals/yd²/ in. depth). Emulsified asphalt prime can be applied at the same rate as emulsified asphalt. Exact application rates are determined by the project engineer under the prevailing conditions at the time the work is done. Application of the MC-250 as a prime should be reserved for open textured bases.

When cutback asphalt is used, the ideal prime rate is the amount of material that the aggregate base will absorb in a 24-hour period. Occasionally, too much cutback asphalt is applied. Even after a normal curing time (24 to 72 hours), some of the asphalt still has not been absorbed into the base. To prevent bleeding of the prime coat up through the asphalt pavement, or creating a slip plane, the excess cutback should be blotted with clean sand or screenings. Blotting involves spreading material over the primed surface and allowing the asphalt prime to be absorbed into the cover material. Paving should be delayed until the excess prime has been properly handled. The area may require a second blotting prior to placement of the HMA. Before the HMA is placed, the prime coated base should be swept lightly with a power broom. Excess sand left in place will prevent a good bond between base course and asphalt layers. Prior to paving, the primed base should be inspected to make sure it has properly cured.

Tack Coats Tack coats are applications of asphalt (usually diluted emulsified asphalt) sprayed on the surface of an existing pavement prior to an overlay. The purpose of a tack coat is to promote the bond between the old and new pavement layers. Tack coats are also used where the HMA comes in contact with the vertical face of curbs, gutters, cold pavement joints and structures.

Tack coat applications are made under the same general weather conditions as HMA paving operations. The roadway surface should be dry and free from dusty material. A power broom is used to sweep the surface prior to tack application. Tack coats are applied just prior to placement of the HMA. If the tack is applied too far out in front, it could lose its tacky characteristic and the surface would require additional tack. The element of time for this to occur is a function of the temperature, wind, surface condition and traffic. If an area is tacked and placement of HMA is not done the same day, the surface should be re-tacked lightly at the beginning of the next paving day.

Before the emulsion *breaks* (the water in the emulsified asphalt begins to evaporate and the asphalt begins to bond with the old pavement surface), a tack-coated surface is slick. Traffic should be kept off the tack coat until no hazardous condition exists, and should be warned of the probability of the emulsion spattering when traffic is permitted on it. The overlay should not be placed until the tack coat has cured to the point where it is tacky to the touch.

The application rate for tack coats is normally 0.25 to 0.70 liters/m² (0.05 to .15 gal./yd²) of diluted emulsion SS-1, SS-1h, CSS-1 or CSS- 1h. The lower application rate is normally used between new or subsequent pavement layers while the intermediate range is for normal pavement conditions, when resurfacing on an existing, relatively smooth asphalt pavement. The upper limit range is recommended for old oxidized, cracked, pocked, or milled asphalt pavements and pcc pavement surfaces. Too little tack coat will not provide a bond, resulting in slippage cracks, and too much can cause slippage between the old and new pavement layers. In addition, too much tack could cause bleeding into the overlay mix compromising mixture stability. The exact application rate should be determined by the project engineer at the time of application.

Although asphalt cements or cutbacks can also be used for tack coats, a diluted, emulsified asphalt (one part water to one part emulsified asphalt) gives the best results for the following reasons:

- Diluted, emulsified asphalt provides the additional volume needed for the distributor to function at normal speed when a lower application rate is used.
- Diluted, emulsified asphalt at ambient temperature flows easily from the distributor, allowing a more uniform application of the tack coat.

When applying either prime coats or tack coats, care must be taken to prevent asphalt overspray onto curbs, gutters, bridge decks, guardrails, or passing automobiles. A wind guard and end nozzles at the appropriate end of the spray bar are recommended to prevent over-spray.

The Asphalt Distributor Prime coats and tack coats are usually applied by an asphalt distributor. As shown in Figure 5.07, an asphalt distributor is a truck-mounted or trailer-mounted asphalt tank with pumps, spray bars, and appropriate controls for regulating the rate at which asphalt is applied to the surface area. A distributor normally includes an oil or gas fired heating system to maintain the asphalt material at the proper application temperature, and a hand-held spray attachment for applying asphalt to areas inaccessible to the spray bars. The heating system is not normally used for emulsions.

When the distributor is not in use, a circulation pumping system keeps the asphalt in motion to prevent the asphalt from solidifying and clogging the spray bar and nozzles.

Medium-curing cutback asphalt, which is usually applied at an elevated temperature, should not be put into a distributor that previously contained an emulsion, unless it is certain that no water remains in the distributor.

Spray Bar Adjustment For normal use, the spray bar of the distributor should be adjusted so that the vertical axes of the nozzles are perpendicular to the roadway. Nozzles should also be set at an angle of 15 to 30 degrees to the horizontal axis of the spray-bar (Figure 5.08) to prevent the fan-shaped spray patterns of each nozzle from interfering with one another. Each nozzle should be set at the same angle.

Another key spray bar adjustment that is essential for uniform prime or tack coat coverage is adjustment of the spray bar height. As Figure 5.09 shows, the fan-shaped spray patterns from the nozzles overlap to different degrees, depending on the distance between the spray-bar and the surface to be covered. The spray bar should be set high enough above the roadway for the surface to receive a single, double or triple coverage. This height will vary according to the nozzle spacing of the spray bar. On some distributors, as asphalt is sprayed and the load lightens, the truck's rear springs rise, raising the distributor and changing the height of the spray bar. Mechanical devices are usually available that automatically correct the spray bar height as this change occurs.

Figure 5.07 A Typical Asphalt Distributor (Courtesy of Etnyre)

The importance of uniform application of asphalt (prime and tack) is essential. Transverse spread (application) should be allowed to vary no more than 15 percent. The longitudinal spread should not vary more than 10 percent. To ensure the correct application, the distributor must be calibrated before it is used. Then the transverse and longitudinal spread rate variations should be checked periodically to determine if the distributor is operating within these limits. These spread rate variations can be checked in the field by using the procedures for checking the application rate of distributors shown in ASTM D2995.

Distributor Controls There are three types of control devices common to most distributors. They are a valve system that controls the flow of asphalt, a pump tachometer or pressure gauge that registers pump output, and a bitumeter with odometer that indicates both the speed of the distributor in number of meters per minute and the total distance traveled by the distributor. All three controls are essential for measuring the quantity of asphalt being applied to the road surface.

The bitumeter should be checked periodically to be certain it is accurately registering the distributor's speed during spraying operations. A buildup of asphalt on the bitumeter wheel is a major cause of bitumeter error. The wheel should therefore be kept clean.

A bitumeter test is conducted on a straight and level length of road. A distance of 175 to 350meters (500 to 1000 feet) is marked off. The distributor is then driven at a constant speed over the marked distance and the trip is timed with a stopwatch. The time elapsed on the stopwatch is used to calculate the distributor's speed in meters (feet) per minute. This speed is com-

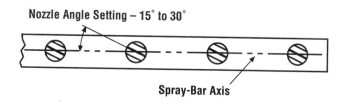

Nozzle Angle Setting – 15° to 30°

Spray-Bar Axis

Note. On occasion, some operators will set end nozzles at a different angle (60 to 90 degrees with respect to the spray-bar) in an attempt to obtain a good edge. This practice should NOT be permitted as it will produce a fat streak on the edge and rob the adjacent spray fan of the lap from this nozzle. A curtain on the end of the bar or a special end-nozzle with all nozzles set at the same angle will provide more uniform coverage and make a better edge.

Figure 5.08 Proper Setting of Nozzles

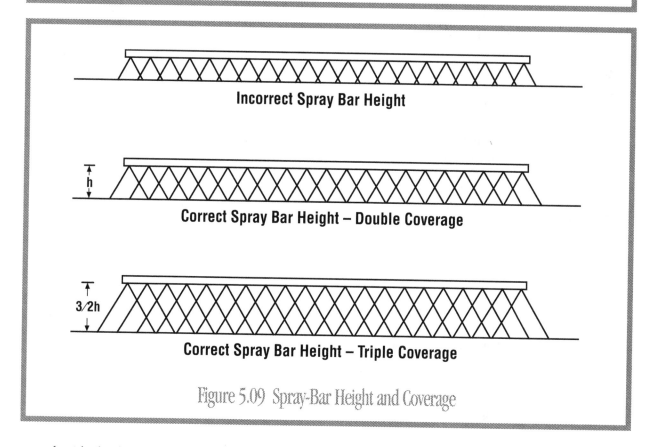

Incorrect Spray Bar Height

h

Correct Spray Bar Height – Double Coverage

3/2h

Correct Spray Bar Height – Triple Coverage

Figure 5.09 Spray-Bar Height and Coverage

pared with the bitumeter dial reading recorded during the test run. The procedure is repeated, with the distributor driven at a different speed on each run. Discrepancies between the reading on the bitumeter dial and the speeds calculated from elapsed times on the stopwatch are noted and are used as correcting factors when spraying operations begin.

Modern distributors are equipped with computer controls. The computer controls the output regardless of the variation in travel speed. Certain inputs to the computer are necessary for proper operation. These include material characteristics, spray bar length, and nozzle size.

Measuring Asphalt Asphalt used for prime coats or tack coats is usually paid for by the liter (or gallon). This means that the contents of each distributor load must be measured before and after spray applications. The difference between the first measurement and the second indicates the amount of material applied to the road. Some distributors have flow gauges that register the amount of asphalt pumped. These gauges must be zeroed (set to zero) before spraying begins and must be read immediately after spraying is completed.

Some distributors are equipped with volume gauges while others require a measuring stick furnished by the manufacturer. They are usually marked in increments of 100 or 200 liters (25 or 50 gallons) depending on the tank size.

When measuring the amount of asphalt in a distributor, it is important to also take the temperature of the material. An accurate asphalt temperature is required to ascertain that the asphalt is at the temperature specified for spraying, and provide the necessary information for making temperature-volume corrections for application and payment. (See Chapter 2, Materials for additional information.)

Calculating Load Coverage It is important to know how much linear distance of roadway can be covered by the asphalt contained in the distributor. The "length of spread" of a distributor load is calculated in U.S. Customary units using the following formula:

$$L = \frac{9T}{WR}$$

Where:

L = Length of spread in feet
T = Total gallons in distributor
W = Sprayed width of roadway in feet
R = Rate of application in gallons per square yard

In metric units, the length of spread is calculated by the following formula:

$$L = \frac{T}{WR}$$

Where:

L = Length of spread in meters
T = Total liters in distributor
W = Sprayed width of roadway in meters
R = Rate of application in liters per square meter

Suggested Spraying Temperatures Table 5.01 shows suggested spraying temperatures for various types and grades of asphalt commonly used for tack coat and prime coat.

Paving Equipment

Paving operations include delivery of HMA to the job site, placement of the mixture on the roadway in compliance with grade and section, and compaction of the mixture to its target density. The delivery and placement of HMA is discussed in this chapter. Compaction is discussed in Chapter 6.

Table 5.01 Spraying Temperature Range for Prime and Tack Coat

Type and Grade of Asphalt	Temperature Range	
	°C	°F
SS-1 SS-1h	20-70	70-160
CSS-1 CSS-1h	20-70	70-160
MC-30*	30+	85+
MC-70*	50+	120+
MC-250*	75+	165+

* Application temperatures may, in cases, be above the flash points of some materials. Care must be taken to prevent fire or explosion.

➤➤ Pavers

Pavers are self-propelled machines designed to place and initially compact an asphalt mixture on the roadway to a specific depth (see Figure 5.10).

The two major parts of a typical paver are the tractor unit and the screed unit (Figure 5.11). Pavers can be equipped with a single lane or full pavement width screed. Full width paver screeds are capable of placing a mat up to a width of 9 meters (30 feet) or more.

Tractor Unit The tractor unit provides moving power for paver wheels or tracks and for all powered machinery on the paver. The tractor unit includes the receiving hopper, feed conveyor, flow control gates, distributing augers (or spreading screws), power plant (engine), transmissions, dual controls, and operator's seat.

In operation, the tractor power plant propels the paver, pulls the screed unit, and provides power to the other components through various transfer devices. HMA is deposited in the hopper, where it is carried by the feed conveyor system through the flow control gates to the distributing augers. The augers distribute the mix evenly across the full width of the paver screed for uniform placement onto the road surface. These operations are controlled by the paver operator by means of dual controls within easy reach of the operator's seat.

The most common method of loading HMA into the paver is for haul trucks to back into the front of the paver and unload the mix directly into the paver hopper. Alternate procedures are often used to help facilitate the operation in achieving all specification requirements of line, grade, and density. One of these alternatives is a windrow elevator, which picks the mix up from a windrow in front of the paver and feeds it to the paver hopper. A Material Transfer Vehicle (MTV) also provides uniform mix delivery to the hopper. These alternatives help provide uniformity of paver speed and auger feeds that translate into smooth pavement with uniform density, thickness and texture.

To ensure that the paver functions properly, several items should be checked prior to commencement of paving:

- *Tires or Tracks* – If the paver is equipped with pneumatic tires, tire condition and air pressure must be checked. It is particularly important for the tire pressure to be the same on both sides of the paver. If the paver moves on tracks (crawlers), they should be checked to be certain they are snug but not tight, and drive sprockets should be checked for excessive wear. Low tire pressure or loose crawlers can cause unnecessary movement of the

Figure 5.10 A Typical Paver (Courtesy of Cedarapids)

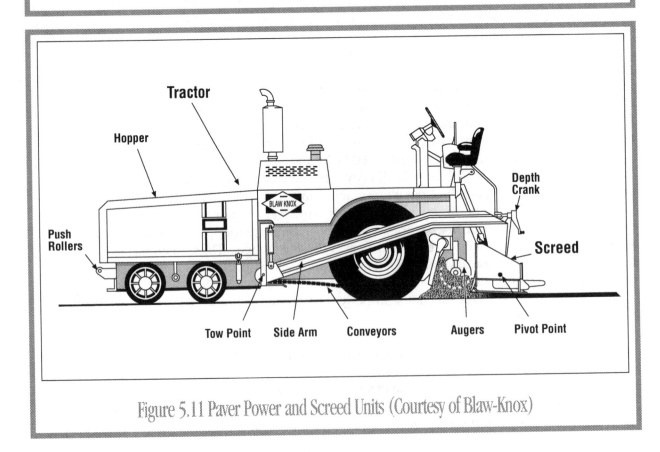

Figure 5.11 Paver Power and Screed Units (Courtesy of Blaw-Knox)

PLACING HOT MIX ASPHALT

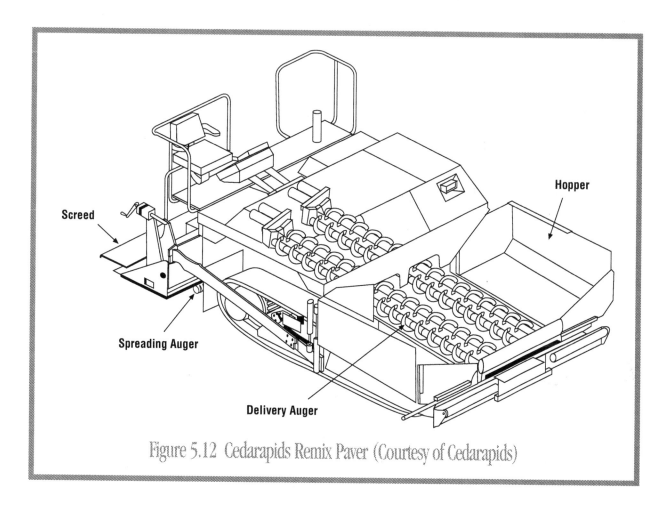

Figure 5.12 Cedarapids Remix Paver (Courtesy of Cedarapids)

paver, which is transmitted to the screed unit, resulting in an uneven pavement surface. There should be no buildup of material on tires or tracks.

- *Hopper, Flow Gates and Auger* – The paver hopper, slats on the feed conveyor, flow gates, and the augers should be checked for excessive wear and observed to be certain they are operating properly. Necessary adjustments should be made by the contractor to ensure that these components are functioning as designed and are able to deliver a smooth flow of mixture from the hopper to the roadway. This includes adjustments to any automatic feed controls.

    The speed of the conveyor and the opening of the control gates at the back of the hopper should be adjusted by the operator so that just enough mixture is being delivered to the augers to keep them operating about 85 percent of the time. This will allow a uniform quantity of mix to be maintained in front of the screed. If additional mix is required to allow an increase in the thickness being placed, the flow control gates should be adjusted. Augers should be kept about three-quarters full of mixture during paver operations.

- *Hopper Feeder Augers* – Some pavers are equipped with remix feeder (delivery) augers (Figure 5.12) to supply mixture to the cross (spreading) augers. The remix feeder augers are comprised of twin, parallel, counter-rotating augers controlled by individual hydraulic motors. The augers are constructed in the lower portion of the hopper deck. The twin augers feed the mixture to the cross augers, eliminating the typical feed slats and the flow gates. The speed of the augers controls the rate at which the material is being received by

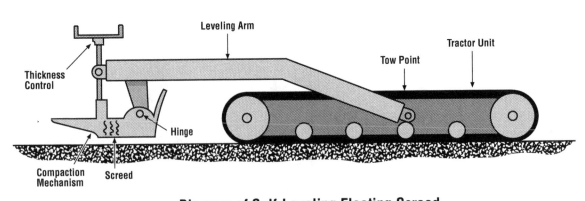

Diagram of Self-Leveling Floating Screed

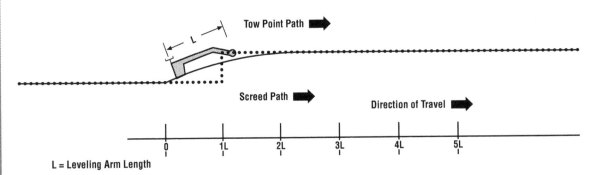

Side View of Screed Path

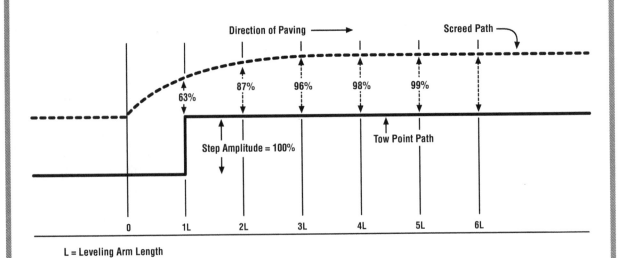

Detail of Screed Path

Figure 5.13 Self-Leveling Floating Screed (Courtesy of Barber Greene)

the cross augers. The speed of each set of feeder augers is controlled independently to maintain the proper level of material in the auger box. This type of auger remix system has had success in controlling segregation. Other pavers have transverse re-mixing paddles installed for the purpose of remixing to control segregation.

- *Leveling Arms* – The leveling arms (pull arms) are the side arm connections from the tow point to the hinge point on the screed. The length varies with paver types and sizes. The length, referred to as the towlength, is important to the leveling effect of the paver. As a general rule, any change in the screed angle will take 5 to 7 towlengths for restoration in the balance of forces and consequent attainment of the desired change in thickness of the HMA layer. The leveling effect with respect to the leveling arm length is shown in Figure 5.13. Note that the step amplitude for the tow point path is completed in one tow length, but the screed path is only at 63% correction at one tow length. The leveling effect is reduced when automatic grade controls are used, depending on where the grade controls are attached to the leveling arms. This may vary with the type of controls used, but most grade controls will optimize at approximately two-thirds the distance from the tow point to the screed. In this case the automatics will reduce the leveling distance to two-thirds the normal distance.

Screed Unit The screed unit has two major functions. It strikes off the mix in a manner that meets specifications for thickness and smoothness and it provides initial compaction of the mixture. A typical screed unit comprises the following:

- Screed pull arms
- Screed plate
- Heating unit
- Tamping bars or vibratory attachments (or a combination of both)
- Controls

In operation, the screed is pulled along behind the tractor unit. The screed pull arms are pivoted, which permits the screed to have a floating action as it travels along the road. As the tractor unit pulls the screed into the mix, the screed seeks the level at which the path of the screed is parallel to the direction of pull. At this level all of the forces acting on the screed are in balance as the paver moves down the road. The screed plate smooths the surface of the mix-

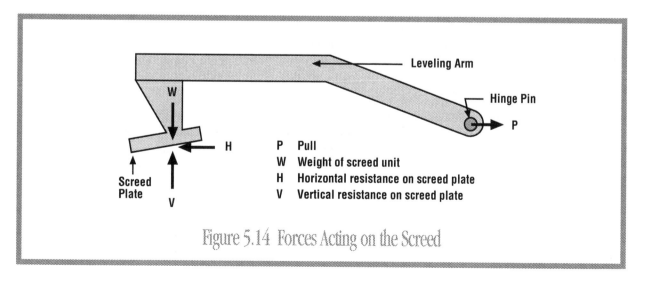

Figure 5.14 Forces Acting on the Screed

ture leaving the mat at a depth that conforms to job specifications. Mat thickness and crown shape are regulated by screed controls. Tamping bars or vibratory attachments then compact the mat slightly in preparation for rolling.

The screed acts automatically toward maintaining position and tends to bring all the forces acting on it into balance. These forces, shown in Figure 5.14, include:

- Forward pull by the tractor (P).
- The force of the material in the augers and moving against the screed (H).
- The downward force of the screed's own weight (W).
- The upward (lifting) force of materials being crowded under the screed (V).

Attaining proper mat thickness is a matter of balancing these forces with one another. For example:

- To maintain forward motion of the screed, force P must be greater than force H.
- To increase the thickness of the mat, the leading edge of the screed plate must be tilted up so that more material is crowded under the screed plate. The screed will rise until the finished surface is again in a plane parallel to the direction of pull. Force V will decrease at this point and be balanced by Force W.
- To reduce mat thickness, the screed plate is tilted downward so that less material crowds under the screed plate.

The amount and condition of material leaving the auger can change the equilibrium of the four forces. Excessive flow of material increases force H. A cold, stiff mix will increase H and to some extent V. An excessively hot, fluid mix decreases H and V. Stopping and starting the paver also causes changes in equilibrium among the forces. The key to controlling the action of the screed is to maintain, in a uniform manner, those forces acting on the screed.

The leveling and compacting function of the screed provides the best results when:

- Forward speed of the paver does not vary.
- Supply of material to the receiving hopper is adequate and constant.
- Distribution of material across the front of the screed is uniform and at a constant level.
- Variables that affect screed leveling are kept under control.
- Compacting forces at the screed are maintained at a constant level.

Therefore, the secrets of good paver operations are balance and uniformity. When balance and uniformity are attained, the screed path follows the paver in a plane parallel to that of the pivot point. As the paver passes over an irregularity, the pivot point rises. The screed begins to rise also. However, because it reacts to changes in elevation more slowly than the pivot point does, the screed rises very little, maintaining the plane of the mat surface over the irregularity. The same is not true of long irregularities (i.e., longer than several lengths of the paver). Grade line irregularities of this type should be corrected prior to placing surface courses with the paver. A fixed stringline would be required to correct this type of grade irregularity.

Screeds with tamping bars and vibratory mechanisms are designed to strike off and compact the mixture slightly as it is placed. Other screeds have a pre-strike-off plate in front of the screed to strike off excess mix as the screed is pulled through the mix. There are two purposes to this screed action. It achieves maximum leveling of the mat surface, and it ensures that minimum distortion of the mat surface will occur with subsequent rolling. Because various screed compaction systems function differently, they are discussed separately below.

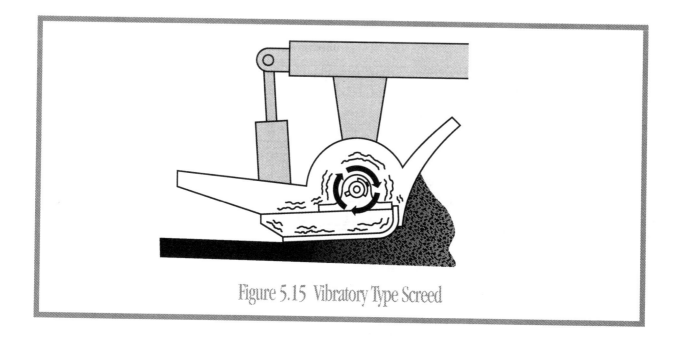

Figure 5.15 Vibratory Type Screed

Tamping Bar Types – Tamping bar-type screed compactors compact the mix, strike off the excess thickness, and tuck the material under the screed plate for leveling. The tamper bar has two faces: a beveled face on the front that compacts the material as the screed is pulled forward; and a horizontal face that imparts some compaction, but primarily strikes off excess material so that the screed can ride smoothly over the mat being laid.

The adjustment that limits the range of downward travel of the tamping bar is the single most important adjustment affecting the appearance of the finished mat. At the bottom of its stroke, the horizontal face should extend 0.4mm (1/64 in.) below the level of the screed plate. If the bar extends down too far, mix builds up on the screed face, which tends to scuff the surface of the mix being placed. In addition, the tamping bar will lift the screed slightly on each downward stroke, often causing a rippling of the mat surface.

If the horizontal face of the tamping bar is adjusted too high (either by poor adjustment or due to wear of the bottom of the horizontal face), the bar does not strike off excess mix from the mat. Consequently, the screed plate begins to strike off the material, which results in surface pitting of the mix being placed, as the leading edge of the screed plate drags the coarser aggregate forward. Therefore, the tamper bars should always be checked before operating the paver. They should be adjusted by the contractor if necessary, and replaced before they approach knife-edge thinness.

Vibrating Types – The operation of vibratory screeds is similar to that of tamping screeds, except that the compactive force is generated either by electric vibrators, rotating shafts with eccentric weights, or hydraulic motors (Figure 5.15). On some pavers both the frequency (number of vibrations per minute) and the amplitude (range of motion) of the vibrators can be adjusted. On others, the frequency remains constant and only the amplitude can be adjusted.

Frequency and amplitude must be set in accordance with the type of paver, the thickness of the mat, the speed of the paver, and the characteristics of the mixture being placed. Once set, the frequency and amplitude do not normally need adjustment until mat thickness or mix characteristics change.

Some vibratory screeds require a pre-strike-off unit. This is a rounded mold board that controls the amount of mix passing under the screed. It is normally set at approximately 3mm (1/8 in.) to 12.5 mm (1/2 in.) above the bottom of the screed, depending on the type of screed.

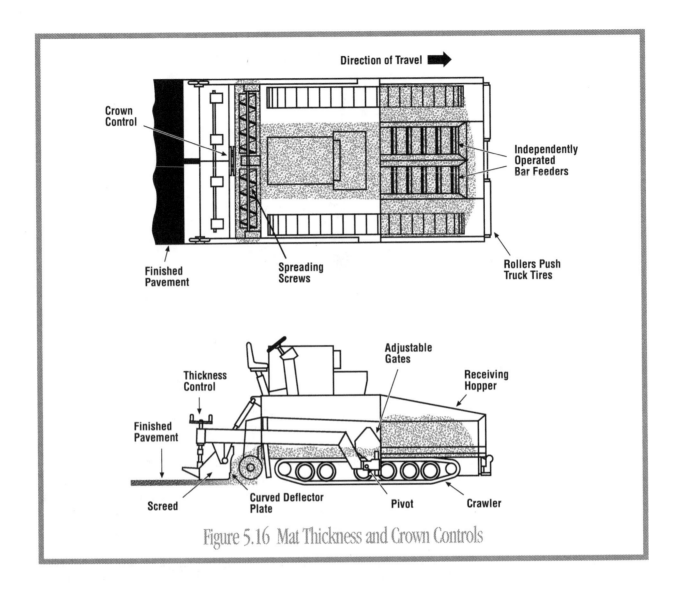

Direction of Travel ➡

Crown Control

Independently Operated Bar Feeders

Finished Pavement

Spreading Screws

Rollers Push Truck Tires

Adjustable Gates

Receiving Hopper

Thickness Control

Finished Pavement

Screed

Curved Deflector Plate

Pivot

Crawler

Figure 5.16 Mat Thickness and Crown Controls

Screed Controls In operating the screed, three types of controls are essential: mat thickness, crown formed in the mat, and transverse slope. All functions are regulated by the controls – the temperature of the mix, and the speed of the paver (Figure 5.16).

Screed control adjustments made to the screed take time to go into effect. It is important that after such adjustment of the thickness controls, the paver be allowed to travel far enough for the correction to be completed before another adjustment is made. Excessive adjustment or over-use of thickness controls is one of the principal contributors to poor pavement smoothness.

The condition of the screed unit is important for placing a high quality mat. Wear points should be checked to be sure that the screed control linkage is snug. Screed plates should also be checked regularly for signs of wear such as pitting and warping. The plates should always be properly adjusted by the contractor before paving begins. Both the leading and trailing edges of the screed have a crown adjustment. The leading edge should always have slightly more crown than the trailing edge to provide a smooth flow of material under the screed. Too much lead crown, however, results in an open texture along the edges of the mat, and too little results in open texture in the center. Crown adjustments may be made independently or simultaneously during the paving operation.

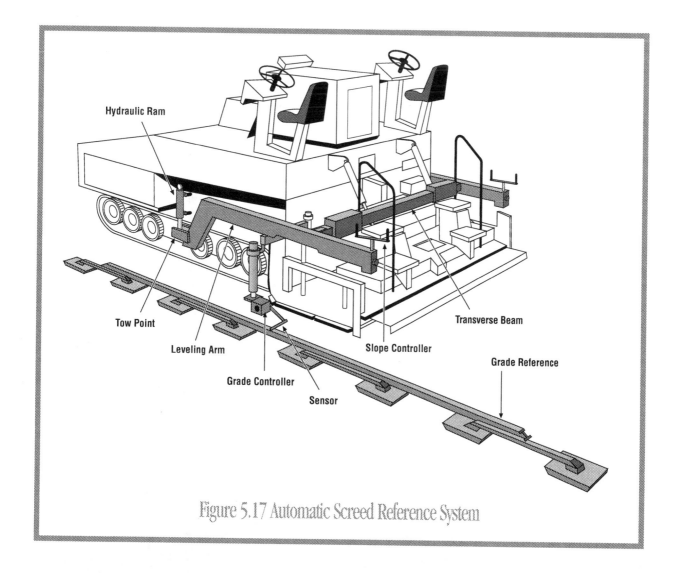

Figure 5.17 Automatic Screed Reference System

Automatic Screed Controls The screed controls just mentioned must be adjusted by the screed operator as paving progresses. Automatic screed controls, however, are designed to adjust automatically to place a uniform mat of desired thickness, grade and shape (Figure 5.17).

There are three basic types of sensing devices used with the automatic screed control system: (1) The wand, (2) the ultra sonic system, and (3) the laser tracer device.

- *Wand sensor* – The wand sensor is depicted in Figure 5.17 as the "grade controller and sensor." It is run on either a fixed stringline or a traveling stringline (also called a ski). The ski simply extends the wheelbase of the paver. The minimum ski length is normally about 7.5meters (25 feet). The more effective ski lengths are from 12 to 18 meters (40 to 60 feet). The traveling stringline is normally preferred on the final course for maximum smoothness. When the grade reference is a fixed stringline the wand is often the preferred method, but other systems can be used equally effectively.

- *Ultra Sonic sensor* – This sensor is mounted near the middle of the pull arm with the sensor face pointed directly at the grade reference. Its operation is based on sonar echoes that are bounced back to the receiver. A sound pulse is bounced off a target reference as an

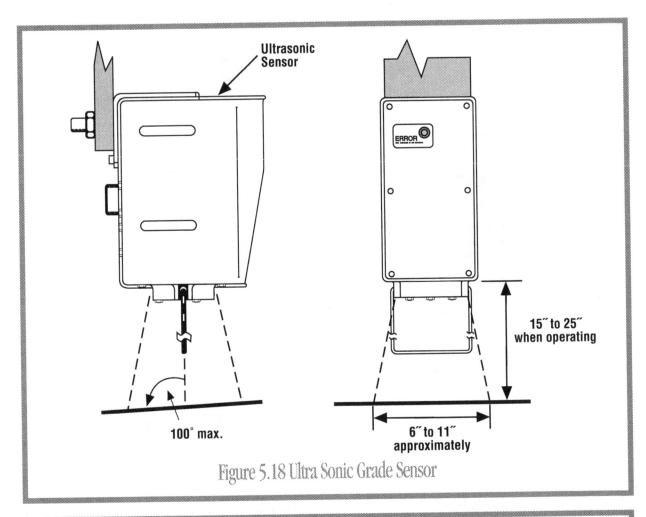

Figure 5.18 Ultra Sonic Grade Sensor

Figure 5.19 Laser Grade Sensor

echo and received by the sensor, and a timer calculates the distance based on the time and speed of sound. The reference with this device can be a stringline, the existing surface, a previously placed lane, or a curb line (See Figure 5.18).

- *Laser sensor* – This device is mounted on the pull arm approximately two-thirds the distance from the tow point to the screed hinge point. This system will follow a stringline, the existing surface, the previous surface, or the curb line. In setting up this system a stringline can be erected that will include roll down factors for true grade. The roll down is estimated to be about 25 percent of the uncompacted mat thickness. To calculate the exact position of the stringline, a survey crew is used to determine the existing grade at approximate intervals of 15meters (50 feet). The existing grade is subtracted from the theoretical grade for calculation of lift thickness. A roll down factor of 25 percent of this thickness is added for the stringline grade. Once the stringline is erected, intermediate points of support may be placed under the stringline, especially on curves or in sudden changes of grade. The laser device incorporates the use of a fan laser and CCD camera. The CCD camera is tuned to see the laser beam. When the stringline breaks the laser beam, the camera shows exactly where it breaks. It then counts the pixel rows down to the break, the vertical position is calculated from this data, and a correction is made to the height of the leveling arm. This system can be operated as a dual system on both sides of the paver or on a single side. Figure 5.19 shows a laser grade sensor.

Automatic screed controls can be used in several different ways, but all automatic screed control operations require a reference system for the automatic system to follow. This reference can be the base on which the HMA is being placed, the lane next to the material being placed, or a stringline.

In addition to reference systems such as those described in the preceding paragraphs, automatic screed controls can also follow traveling reference systems. In such a traveling reference system, a ski is attached to a control arm, and this ski notes changes in the base contours and adjusts the screed automatically to compensate.

To maintain proper transverse grade, automatic screed controls use a system attached to a beam running between the two screed pull arms. A pendulum in the slope control housing moves side-to-side with changes in the transverse grade of the roadway, triggering necessary adjustments in the slope control mechanism.

Automatic control systems have several advantages over manually controlled screed systems. Some of them are:

- Automatic controls can compensate for changes in grade and slope more quickly than a screed operator.
- Automatic controls help disassociate the screed from erratic vertical movement of the tractor unit.
- Automatic controls adjust the screed towpoints to enable the screed to follow a path parallel to the grade and slope of the reference system, which may be different from the path plane of the tractor unit.

Selecting a Reference System
The reference system that is best for the job depends on the existing pavement surface and the amount of maneuvering back and forth required by the project conditions. Some of the factors to consider are: the condition of the surface on which the mat is to be placed, the degree of precision required in the grade and slope of the finished pavement, the thickness of the mat, and the amount of material available for the project. If the

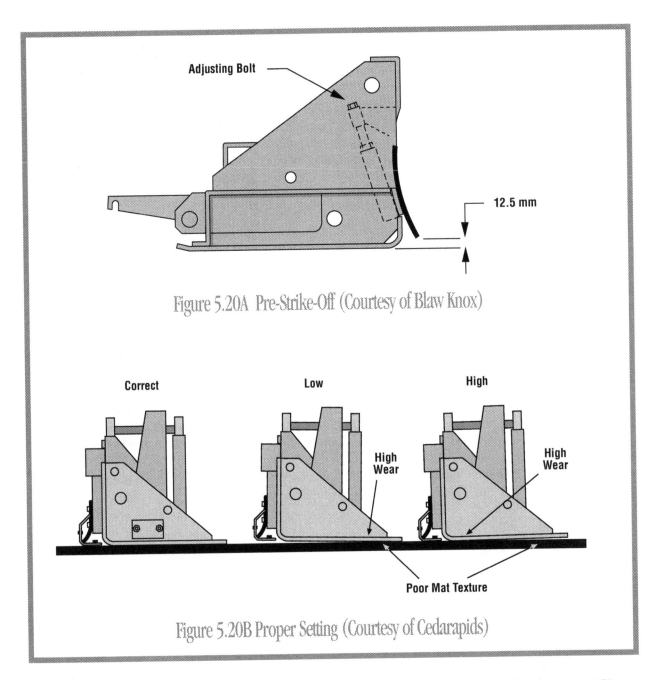

Adjusting Bolt

12.5 mm

Figure 5.20A Pre-Strike-Off (Courtesy of Blaw Knox)

Correct   **Low**   **High**

High
Wear

High
Wear

Poor Mat Texture

Figure 5.20B Proper Setting (Courtesy of Cedarapids)

surface on which the mat is to be placed has an appropriate longitudinal grade along its PGL, but an unsatisfactory transverse grade, a traveling reference can provide the desired mat thickness at the PGL. The transverse slope control can be used to establish the adjacent grade. In using the transverse slope control, no specific roll down factors can be applied to grade calculations for slope transfer.

If the longitudinal grade is erratic, a stringline should be placed to ensure a proper longitudinal grade and transverse cross slope. If the existing surface has a good profile both longitudinally and transversely, the traveling stringline can be used with a minimum ski length. The self-leveling ability of the screed may be sufficient for scratch or base course mixtures. The final decision should be made in view of the smoothness specification requirements and the nature of the project.

Control Systems Either a single or dual grade control system can be used to control the paver.

- *Single Grade Control System* - A single grade control is set for one side of the paver and the opposite side is controlled from the grade side by using the cross slope device. In situations where the paving width is nominal and the existing surface is fairly uniform this will work quite well. Roll-down factors can be put into the grade control side and will be transferred to the opposite side as equal factors, which may or may not be equal.

- *Dual Grade Control System* - In this system, as its name implies, the paver is using two different grade control systems, one on each side of the paver. This can include any one of the systems previously described. It may involve a wand with a fixed or traveling stringline on one side with the laser, sonic, or joint matcher on the other side.

Setting the Screed The older pavers are manufactured with a pre-strike-off blade. This is a cutting blade mounted on the front of the screed. The purpose of the pre-strike-off blade is to slice the mass of material in a plane parallel to the line of pull by the leveling arms as the paver is powered into the material mass. If the paver is equipped with a pre-strike-off, it should be set at 3.2 mm (1/8 in.) to 12.5mm (1/2 in.) above the nose point of the screed (See Figure 5.20A). The screed has forward crown and rear crown, which is set to allow the material to flow evenly, yet be uniformly confined under the screed as it is pulled through the mass. If not properly set, mat texture will be non-uniform and there will be excessive wear on the screed, as shown in Figure 5.20B. Crown adjustment controls are shown in Figure 5.16 for front and rear of the screed.

Starting Blocks Starting blocks equal to 1.25 times the thickness of the uncompacted mat are required to set the thickness and to null the screed. By using starting blocks the grade can be very close at the beginning of the operation. Blocks equal to 25 percent of the uncompacted thickness are used to start from a joint. The 25 percent additional thickness allows for proper roll-down or compaction while maintaining proper grade. Figure 5.21 illustrates the use of starting blocks under the screed.

Various thicknesses of blocks or shims should be available on the project for joints and starting. Projects that contain several different courses of mixtures at different thicknesses will require several different thicknesses of shims. Extended screeds will require multiple shims for each extension area.

Minimum Compacted Thickness The minimum recommended compacted thickness of an HMA course is three times the nominal maximum size aggregate. When the mat falls below this thickness, it pulls, tears, cools rapidly and generally will not be able to achieve the proper density and pavement smoothness.

Screed Heaters The screed is equipped with heaters used to preheat the screed plate at the start of each new paving operation. The heaters are not used to heat the mix during the paving operation. If the screed is not initially heated, the mix will tear and the texture will look open and coarse, as if the mix were too cold.

Screed Accessories Three types of commonly used screed accessories are screed extensions, cut-off shoes, and slope plates. Screed extensions are attachments that widen the screed, allow-

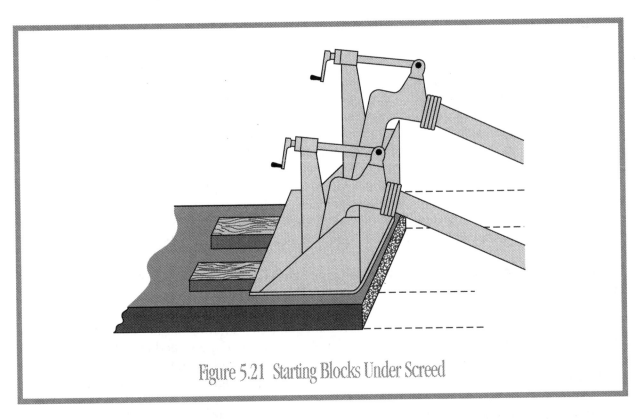

Figure 5.21 Starting Blocks Under Screed

ing the paver to place a wider-than-normal mat. Cut-off shoes have the opposite function. They are metal plates inserted into the screed to reduce the width of the mat being placed. Slope plates are metal plates that are placed at the end of the screed and can be adjusted to place a particular slope on the pavement edge.

Truck Hitches The purpose of a truck hitch on the front of the paver hopper is to keep the dump truck in contact with the paver. If, during dumping, the truck and the paver separate and HMA spills, it must be cleaned up before the paver passes over it or it will cause roughness in the mat.

There are two types of truck hitches in common use: One type utilizes an extension that reaches under the truck and hooks onto the truck's rear axle. The other system uses retractable rollers that are attached to the truck push bar and grip the outer side of the truck's rear wheels. They revolve with the wheels while the truck dumps its load into the hopper.

Pivoted Truck Push-Rollers The pivoted push-roller is a device mounted on the front of the paver that adjusts when alignment between the truck and the paver is uneven. The device reduces the uneven force exerted on the paver by the misaligned truck, minimizing interference in the steering of both vehicles.

►►Haul Trucks

HMA is delivered to the job site by trucks. The particular trucks used for delivery of HMA should meet all specification requirements of size, configuration and tarp standards. Any release agents used to treat truck beds should be approved by the agency. The basic information for truck requirements is shown in Figure 5.22

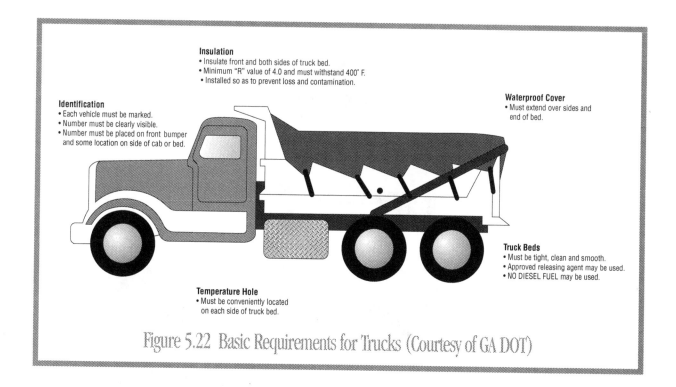

Insulation
• Insulate front and both sides of truck bed.
• Minimum "R" value of 4.0 and must withstand 400° F.
• Installed so as to prevent loss and contamination.

Waterproof Cover
• Must extend over sides and end of bed.

Identification
• Each vehicle must be marked.
• Number must be clearly visible.
• Number must be placed on front bumper and some location on side of cab or bed.

Truck Beds
• Must be tight, clean and smooth.
• Approved releasing agent may be used.
• NO DIESEL FUEL may be used.

Temperature Hole
• Must be conveniently located on each side of truck bed.

Figure 5.22 Basic Requirements for Trucks (Courtesy of GA DOT)

General Information Various types of trucks are used to deliver HMA to the job site. The two most common types are end-dump trucks and bottom-dump trucks (Figure 5.23). Details regarding use of each type are presented in later sections. Presented below is general information pertinent to all types of trucks used for hauling HMA.

The number of trucks required on the project is determined by many factors: the mix production rate at the plant, the length of the haul, the type of traffic encountered, and the expected time needed for unloading. The number of specific trucks required to keep the paver moving can be estimated using the round trip time divided by the load time, plus one.

When delivering HMA to the paver hopper, the truck bed should be raised to a height of 0.9 to 1.2 meters (3 to 4 feet) prior to releasing the tailgate. This will place a surge against the tailgate so that, when released, the mixture will flood the hopper. When the tailgate is released without raising the bed, segregation occurs across the hopper. Each time the bed is raised higher, the mixture should move in mass toward the tailgate. The hopper is then flooded each time the height of the bed is increased.

Condition of Haul Trucks Trucks must have metal beds, and the beds must be clean, smooth and free of holes. All trucks must meet minimum safety criteria. Each truck must be clearly numbered for easy identification and must be equipped with a tarpaulin.

The truck bed must be cleaned of foreign material and hardened asphalt before loading. It is then lightly coated with a release agent (lubricant) that assists in preventing fresh hot mix asphalt from sticking to the surfaces of the bed. After the bed is coated, any excess release agent must be drained from the bed. Before loading, the truck must also be weighed to establish a tare weight (unloaded weight). The tare weight is later subtracted from the loaded weight of the truck to determine the weight of HMA the truck is hauling.

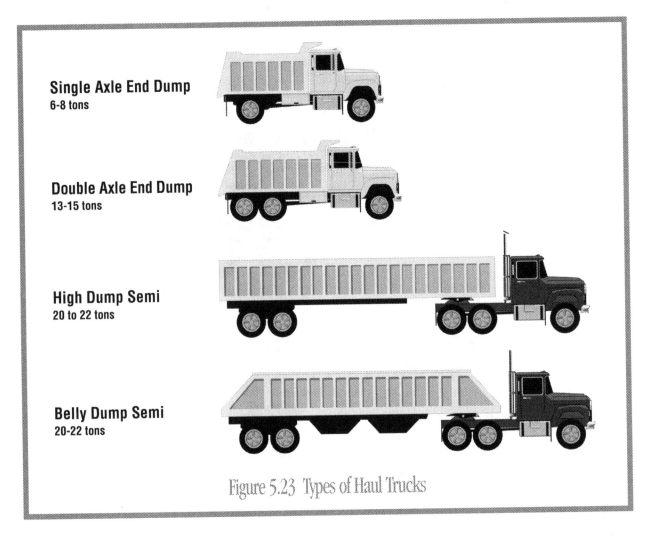

Single Axle End Dump
6-8 tons

Double Axle End Dump
13-15 tons

High Dump Semi
20 to 22 tons

Belly Dump Semi
20-22 tons

Figure 5.23 Types of Haul Trucks

Types of Haul Trucks Each type of truck used for HMA delivery must have certain physical features in order to properly haul and discharge into the paver. Below are listed a few guidelines for the two most common types of trucks.

- *End-Dump Trucks* – An end-dump truck must be inspected to be certain the rear of the bed overhangs the rear wheels enough to discharge mix into the paver hopper. If it does not, an apron with side plates must be added to increase the overhang and prevent spillage of mix in front of the paver.

    The bed must be of a size that will fit into the hopper without pressing down on the paver. Such pressure could result in damage to the paver wings and corner flaps. The hydraulic system for the truck-bed hoist should be frequently inspected to guard against hydraulic fluid leakage. Such leakage on the roadway surface will strip the asphalt cement from the aggregate. If enough of this oil is spilled, the mix will absorb it and become unstable, requiring removal and replacement.

    Tarpaulins should be pulled over the mixture in the truck beds during hauling in cool weather or on long hauls to protect the mixture from excessive cooling. A cool mix forms lumps and a crust over its surface. If a tarp is used, care must be taken to be sure it is securely fastened to the top of the truck bed so that cold air cannot funnel under it (see Figure 5.22).

Figure 5.24 Material Transfer Vehicle (Courtesy of ROADTEC)

During delivery, the "dump" person must direct the truck squarely with respect to the paver, stopping a few feet short of the hopper. The paver will then pull into the truck, picking it up and locking on for a smooth forward push. Backing the truck against the paver can force the screed back into the mat, leaving a bump in the pavement even after the mat is rolled.

- *Bottom-Dump Trucks* – If bottom-dump trucks are used, the hot mix is dumped in a windrow. A pick-up device is then necessary to place the HMA in the hopper.

    There are two common methods for unloading bottom-dump trucks. One method involves the use of a spreader box designed to operate under the gates of the truck. The amount of material placed in the windrow is governed by the width of the spreader box opening. The disadvantage of this method is that the spreader box can restrict the amount of material to less than the required amount.

    The second method utilizes chains to open the dump gate. This is the most commonly used procedure. Automatic devices are available for controlling gate openings; however, their use is somewhat limited because of the additional cost.

    Variations in the size of the windrow and irregularities in the surface on which the material is to be placed will cause variations in the amount of material fed to the paver hopper. This often causes variations in the finished surface. It is therefore essential that the windrow deposited by the truck be as uniform as possible. If the windrow is deficient in size, material can be added to it to keep the paver from starving. If the windrow contains too much mix, a short gap in depositing with the next truck will compensate for the excess. The windrow length must also be controlled, particularly in cool weather. Windrowed material will cool below spreading and compaction temperatures in cool weather, particularly when delay occurs because of paver malfunction. To prevent excessive cooling of the mix in cold weather, the limit of the windrow should be no more than one truckload ahead of the pick up machine.

- *Material Transfer Vehicles* - A Material Transfer Vehicle (MTV) can be used to transfer material from the truck to the paver without segregation and, at the same time, offer a uniform delivery system. The MTV device will enable the paver to operate at a uniform speed and provide a uniform head of materials flow to the auger system. The throats on the paver can be adjusted to provide constant auger action, thus eliminating the center lane crack often seen as a result of the augers not turning. Some MTVs can hold as many as 45 metric tons (50 tons) of mix. Some models also provide the windrow maker if that is desirable. The MTVs offer the best assurance for uniform materials delivery, constant paver speed, uniform mix temperature, and elimination of end of truckload segregation. Pavement smoothness, texture and density are critical to pavement performance. The use of the MTV can improve each of these factors. A typical MTV is shown in Figure 5.24

Delivery of Hot Mix Asphalt

➤➤ Load Tickets

Load tickets provide essential records for the control of project operations, quality, and quantity of mix delivered. A typical load ticket is shown in Figure 5.25. Although different systems are used by the various agencies, certain items related to tickets remain generally the same from project to project. Load tickets numbered consecutively are issued at the HMA plant. They show the project number, the origin of the load, date, time loaded, weight of the load, accumulated weight, truck number, and the type of mix. They may also list the temperature of the mix at the time a load leaves the plant. This is done periodically to provide the roadway crew the temperature loss during the travel time.

Several things of importance are contained on these tickets. First, the numbering of the tickets consecutively will show whether or not a truck arrived at the paver in an order different from which it was loaded. This could be due to breakdown of the truck, traffic problems, or something else, but it will give the project personnel an idea of the length of time that the material has been loaded. If the period of time is longer than would be considered normal, the mix should be checked (more thoroughly than usual) for proper temperature. If serious temperature problems are detected, the load should be rejected. It is important for the "Dump" person to collect the load tickets from each truck as the truck is dumping. This provides an accounting of all the trucks that were intended for the project.

➤➤ Visual Inspection of the Mix

During the pre-construction conference, mix quality control procedures should be thoroughly discussed. Although the mix is periodically checked at the plant, there are times when a deficient truckload resulting from a plant malfunction may be inadvertently overlooked. Some deficiencies can be readily noticed by the "dump" person prior to dumping material into the paver. When the temperature is checked or the truck bed is raised, these deficiencies may be readily

FRESHOUR CONSTRUCTION CO., INC

0069636

Ticket Number	69636	
DATE: 09/09/97	TIME: 6:12 AM	

SOLD TO:
AHTD

( ) -

CUSTOMER NO.
01

JOB ORDER NUMBER
12

DELIVERED TO:

TRUCK INFORMATION

NUMBER	LOADS TODAY	TONS TODAY
297	6	107.768

MATERIAL LOADED

SILO or SPEC	CODE	PRODUCT NAME	TONS SOLD TODAY
01	25 MM		656.113

PLANT NO. 01

OPERATOR BOBY

WEIGHT IN TONS

GROSS	TARE	NET
61299	25360	35939
62000	← TRUCK MAXIMUMS →	36640

COMMENTS/DIRECTIONS:

CONTRACT # 060592

CASH SALE

PRICE PER TON	$
MATERIAL COST	$
HAUL COST	$
SALES TAX	$
TOTAL	$

JOB INFORMATION

LOAD NO.	TONS TODAY	TOTAL TONS TO DATE	TONS REQUIRED	LOCATION or JOB NAME
38	656.114	4683.952		

Driver: _____   Rcvd By: _____

Figure 5.25 Load Ticket

---

apparent. Some indications of HMA deficiencies that may require close inspection and possible corrective action are:

- *Blue Smoke* – Blue smoke rising from the mix in the truck or the paver hopper may indicate an overheated batch. The temperature should be checked immediately.

- *Stiff Appearance* – Generally, a load that appears stiff or has an unusually high peak, may be too cool to meet specifications. The temperature should be checked. If it is below the optimum placing temperature, but within the acceptable temperature range, immediate steps should be taken to correct the cause of the low temperature and thereby decrease the possibility of having to waste mix.

- *Mix Slumped in Truck* – Normally the material in the truck is in the shape of a dome. If a load lies flat or nearly flat, it may contain too much asphalt or excessive moisture. Close inspection should be made at once. Excess asphalt also may be detected under the screed as excessive brightness on the mat surface. A mix containing a large amount of coarse aggregate might be mistaken for an over-asphalted mix because of its shiny appearance. Such a mix, however, usually will not slump in the haul truck. If the mixture contains excess moisture, one might see moisture dripping from the truck bed. This is normally the result of internal moisture in the aggregate. In this case the mixing temperature needs to be lowered to the point where this does not happen. This is usually slightly below 150°C (300°F).

• *Lean, Dull Appearance* – A mix that contains too little asphalt can generally be detected immediately in the truck or in the paver hopper by its brown appearance or lack of typical shiny black luster. Absorptive aggregates are most often the cause of under-asphalted mixtures. Excess fine aggregate or bag house fines can cause a mix to be low in asphalt.

• *Excess Moisture* – Excess moisture often appears as lumps or clods in the mix when it is dumped into the hopper of the paver. The mixture would also have a bright shine and popping as if it were boiling. Excessive moisture may also cause the mix to appear and act as though it contains excessive asphalt.

• *Segregation* – Segregation of the aggregates in the mix may occur during paving because of improper handling, or it may have happened at some point prior to the mix reaching the paver. In any case, corrective action should be taken immediately. The cause of the segregation should be corrected at its source. Chapter 8 contains additional information on segregation.

• *Contamination* – Mixes can become contaminated by a number of foreign substances. Trucks that are hauling other materials and are not inspected and cleaned prior to hauling HMA could be contaminated. The contamination can be removed if it is not too extensive; however, a load that has been thoroughly contaminated should be rejected.

• *Bleeding* – While non-petroleum-based agents are normally required for spraying truck beds, diesel fuel may occasionally be used. Excess diesel fuel that collects in the truck bottom can be absorbed by the mix. The fuel dilutes the asphalt and reduces its binding quality. Bleeding may occur if sufficient fuel is combined with the mix. HMA contaminated with diesel fuel should be removed and replaced.

➤➤ *Calculating Paver Yield*

The weight of a truckload is used to check the yield of the paver (length of pavement section per truckload of mix). The starting point is to know how much the mixture weighs after compaction. Once that information is obtained, measurements of the mat placed and some simple calculations can tell whether the paver yield is close to that expected. Following is an example problem in which U.S. Customary units are used:

PROBLEM:

A truck delivers 15 tons (30,000 pounds) of hot mix to the paver. The paver is placing a mat 12½ feet wide and 1½ inch thick (compacted). The mixture has an in-place density of 144 pounds per cubic foot. How long a section (how many linear feet) of pavement should the paver be able to place using the 15 tons?

SOLUTION:

1) A cubic foot of mixture weighs 144 pounds. A square foot of the mat 1 inch thick contains 12 pounds of mix.

$$\frac{144}{12} = 12 \text{ lbs./ft.}^2/\text{per inch of thickness}$$

2) Since the mix is being placed 12½ feet wide, 1½ inches thick, the weight of mix per linear foot of paving is

$$12½ \times 1½ \times 12 \times 1 = 225 \text{ lbs.}$$

3) By dividing the weight of the truck load (30,000 lbs.) by the mixture weight per linear foot (225 lbs.), the number of linear feet of pavement that the paver should place using the load is determined. Therefore:

ANSWER:

$$\frac{30,000 \text{ lbs.}}{225 \text{ lbs./lin. ft.}} = 133⅓ \text{ linear feet}$$

The paver should be able to pave 133⅓ linear feet of pavement with the load delivered by the truck. Taking the calculation one step further, one ton of mix will pave 8.89 feet. This information can be used to check the accumulated total weight from the load tickets against the actual number of feet placed. It can also be used to determine the amount of HMA required to pave a given length of road. And toward the end of the day, it can be used to calculate how much mix is needed to finish a length of roadway.

Placing Procedures

►►Coordinating Plant and Paver

Uniformity of operations is essential in hot mix asphalt paving. Uniform, continuous operation of the paver produces the highest quality pavement.

Paving too fast can result in the paver stopping frequently to wait for trucks to bring more mix. The smoothness of the pavement will suffer when the paver stops and starts up again. The paver speed should be matched to the quantity of HMA being delivered to the project to provide a uniform paver speed.

Obviously, then, it is essential that plant production and paving operations be coordinated. The paver must be continuously supplied with enough mix, and at the same time, trucks should not have to wait a long time to discharge their loads into the paver hopper.

►►Screed Adjustment During Paving

If the mat being placed is uniform and satisfactory in texture, and the thickness is correct, no screed adjustments are required. But when adjustments are required, they should be made in small increments. Time should be allowed between the adjustments to permit the paver screed to complete reaction to the adjustments sequentially.

It is equally important that thickness controls on the screed not be adjusted excessively either in amount or frequency. Each adjustment of the thickness controls results in a change in elevation of the mat surface. Excessive changes in the surface elevation at the edge of the first mat are extremely difficult to match in the companion lane when constructing the longitudinal joint.

➤➤ Width of Spread

Longitudinal edges of successive lifts of mix should not be constructed directly over each other, but offset at least 150 mm (6 in.) on alternate sides of the centerline or base line on succeeding lifts. For example, on a 7.3 m (24-ft.) pavement, the first course (lane) is 3.8 m (12½ ft.) wide and the adjoining lane 3.5 m (11½ ft.) wide. The subsequent lift is then placed at a width of 3.7 m (12 ft.) which places the joint on the center line. This prevents a continuous vertical seam through the completed pavement along the longitudinal joint.

Alignment of the mat is dependent on the accuracy of the guideline provided for the paver operator and his alertness in following it. Attention to this detail is vital to the construction of a satisfactory longitudinal joint, since only a straight edge can be properly matched to make the joint. On a wide roadway, where multiple lanes are being placed, it is generally best to place the lane adjacent to the PGL first, and then match the adjoining lanes to it.

➤➤ Handwork

There are places on many projects where spreading with a paver is either impractical or impossible. In these cases, hand spreading may be required. Placing and spreading by hand should be done very carefully and the material distributed uniformly so there will be no segregation of the mix. When the HMA is dumped in piles, it should be placed far enough ahead of the shovelers to necessitate moving the entire pile. Also, sufficient space should be provided for the workmen to stand on the base rather than on the pile of material being spread. If the asphalt mix is broadcast with shovels, segregation of the mix will result. It is desirable to place the mixture carefully and try to blend the surface texture with the overall project texture. A mixture placed by hand will have a different surface appearance than the same mixture placed by machine.

The material should be deposited from the shovels into small piles and spread with lutes. In the spreading process, all material should be thoroughly loosened and evenly distributed. Any part of the mix that has formed into lumps and does not break down easily should be discarded. After the material has been placed and before rolling starts, the surface should be checked with templates or straightedges and all irregularities corrected.

Inspection of Mat

The crew must be able to identify deficiencies in the finished pavement and understand possible causes of those deficiencies. Table 5.02 lists some common pavement problems and their probable causes. In referring to the table, note that a given deficiency may have several possible causes. In many cases, sampling and testing is the only reliable means for analyzing a pavement problem.

The following sections discuss in detail several of the major items of concern in the placement operation.

➤➤ Temperature of the Mix

The mix temperature is usually established at the plant. It is checked there and also in the trucks; however, it should also be checked behind the paver to establish a uniform temperature for compaction. This is particularly important early in the day when both the surface on which the material is being placed and the air are cool. It must also be checked periodically in conjunction with various compaction phases.

A recommended procedure is to use an infrared digital thermometer, taking several readings across the pavement. Mat temperature will vary in accordance with mat thickness. The mat always cools from the bottom up and from the top down. Mix temperature may require adjustment to accommodate compaction. Consistent mat temperature is essential for uniform compaction of the mat. By monitoring the surface temperature and holding the mat thickness constant, the placement and rolling procedure can remain constant.

➤➤ Pavement Surface Appearance

Surface Texture The texture of the unrolled mat should appear uniformly dense, both transversely and longitudinally. If tearing or open texture appears only at the beginning of paving, it is probably caused by the screed not being heated sufficiently. If irregular texture or tears appear under screed extensions, the alignment of the extension, the tamping bars, or vibrators need to be checked.

- *Tearing or Scuffing* – Tearing often occurs in a mix that is too cold or low in asphalt content and will appear open and coarse. Tearing and scuffing will also result from improper setting of a paver equipped with a tamping bar in the screed unit. The speed of the paver, lift thickness, and frequency of the vibrator screed also may require adjustments to overcome surface irregularities.

- *Texture Irregularities* – A mix containing excess moisture can not be laid properly, and it will have the appearance of a cold mix or an over-asphalted mix. In addition to possibly tearing, the mix will blister. If the moisture can not be eliminated or reduced to an acceptable level, silicones can be used to aid in the placement.

Surface Smoothness Pavement smoothness is adversely affected by a lack of uniformity in the paving operation, improper aggregate gradations, variations in paver speed, improper operation of trucks, poor joint construction practices, and poor grade controls.

- *Lack of Uniformity* – Stopping the paver will cause roughness in the pavement. Every time the paver stops, there is a possibility of the screed leaving a mark on the surface of the mat. If the screed settles into the mix, it causes the automatic sensor to act as if the paver has traveled into a depression. As the paver starts off, the screed lays a thicker mat. This continues until the sensor recognizes the excessive thickness and decreases the slope of the screed. Then a dip is developed until the screed levels out, approximately 9 meters (30 feet) from where the paver stopped.

  Rough pavements also result from non-uniform amounts of material in front of the screed. If there is not enough material in front of the screed, the screed will drop. If there is too much material in front of the screed, it will rise.

- *Segregation* – Excessively coarse aggregate may result in a harsh mix that creates a coarse texture and an uneven surface. Excessive fines may cause a low stability in the mix and permit ripples to form. Segregation may be associated with several factors. The source must be located and corrected.

Table 5.02 Typical Problems and Their Probable Cause (Courtesy of Barber-Greene)

Causes (columns): Cold Mix Temperature; Variation of Mix Temperature; Moisture in Mix; Mix Segregation; Improper Mix Design (Asphalt); Improper Mix Design (Aggregate); Parking Roller on Hot Mat; Reversing or Turning Too Fast of Rollers; Improper Rolling Operation; Improper Base Preparation; Truck Holding Brakes; Trucks Bumping Finisher; Improper Mat Thickness for Maximum Aggregate Size; Sitting Long Period Between Loads; Improper Joint Overlap; Grade Reference Inadequate; Grade Control Wand Bouncing on Reference; Grade Control Hunting (Sensitivity Too High); Grade Control Mounted Incorrectly; Vibrators Running Too Slow; Screed Extensions Installed Incorrectly; Screed Starting Blocks Too Short; Incorrect Nulling of Screed; Kicker Screws Worn Out or Mounted Incorrectly; Feeder Gates Set Incorrectly; Running Hopper Empty Between Loads; Moldboard on Strikeoff Too Low; Cold Screed; Screed Plates Not Tight; Screed Plates Worn Out or Warped; Screed Riding on Lift Cylinders; Excessive Play in Screed Mechanical Connection; Overcorrecting Thickness Control Screws; Too Little Lead Crown in Screed; Too Much Lead Crown in Screed; Finisher Speed Too Fast; Feeder Screws Overloaded; Fluctuating Head of Material.

Problems (rows):
- Wavy Surface – Short Waves (Ripples)
- Wavy Surface – Long Waves
- Tearing of Mat – Full Width
- Tearing of Mat – Center Streak
- Tearing of Mat – Outside Streaks
- Mat Texture – Nonuniform
- Screed Marks
- Screed Not Responding to Correction
- Auger Shadows
- Poor Precompaction
- Poor Longitudinal Joint
- Poor Transverse Joint
- Transverse Cracking (Checking)
- Mat Shoving Under Roller
- Roller Marks
- Poor Mix Compaction

(Matrix cross-referencing each problem with its probable causes. Checks (✓) indicate causes related to the paver; X's indicate other problems to be investigated.)

NOTE: Many times a problem can be caused by more than one item, therefore, it is important that each cause listed be eliminated to assure solving the problem.

Procedure for Using Table

1. Find problem above.
2. Checks indicate causes related to the paver. X's indicate other problems to be investigated.

➤➤Pavement Lane Crown

If there is an open or torn texture at the center of the mat behind the paver, additional lead crown is needed in the front edge of the screed. This forces more mix into the central portion of the screed, closing the texture. If the tears occur on the outer edge, there may be too much crown in the leading edge, forcing too much material in the center and too little at the edges. Reducing the center crown slightly will distribute more material toward the edges and provide a more uniform mat.

➤➤Pavement Geometrics

Geometrics refers to the physical dimensions and shape of the finished pavement including the longitudinal grade, transverse grade, alignment, cross-slope, and pavement thickness. Checking the geometrics of a finished mat begins with knowing the typical section of the pavement. All measurements must be compared with the plans to determine whether the pavement's dimensions and shape are acceptable.

Profilographs and other devices are used to check smoothness, which is usually measured in inches per mile. Transverse grade and width can be checked with stringlines, straight edges, and measuring tapes as well as line-of-sight devices, using grade and elevation markers as reference points.

Before compaction, mat thickness can be determined by using a depth gauge (Figure 5.26), or by extending a straightedge over the edge of the mat and measuring the distance between the straightedge and the previous surface. This loose depth of the mat can be compared with the yield to determine if the correct quantity of material in accordance with the plans and specifications is being placed.

After compaction, measuring with the straightedge can be repeated. In most cases, however, final pavement thickness is measured by coring the pavement. Cores of compacted mixtures are cut from the finished pavement both for measurement of thickness and for density testing (Figure 5.27). Generally, it is easier to measure the depth of the mat by this method.

➤➤Joints

Pavement joints are seams between adjacent mats. The joints in the pavement all too often dictate the life of the pavement. Poor density at joints invites pavement aging by allowing moisture and air into the pavement. There are two types of pavement joints: transverse joints and longitudinal joints.

Transverse Joints A transverse joint occurs at any point where the paver ends work and then resumes work at a subsequent time. Transverse joints are constructed in two steps: (1) The ending of the first lane or width of pavement at the close of work; and (2) The resumption of pavement operations or the start of work on the first day of paving.

Ending a Lane When terminating paving operations at the end of a day's work, the pavement mat must be cut off vertically so that a full depth lift can be placed squarely against it. This requirement can be satisfied by the following procedure:

1) When the paver is placing the last load of the day, it is shifted into low gear as it approaches the location of the proposed joint.

Figure 5.26 Checking Mat Thickness With a Depth Gauge

Figure 5.27 Taking Core Sample of Pavement and Measuring Mat Thickness of Core

PLACING HOT MIX ASPHALT

2) As the hopper empties and the amount of material in the screed chamber decreases below normal operating level, the paver is stopped.

3) The screed is raised and the paver moved out of the way.

4) The mix is then removed from the end of the mat to form a clean, vertical edge.

5) A board or heavy wrapping paper is placed along the edge as shown in Figure 5.28.

6) The material that was removed in Step 4 is replaced and used to form a taper, or ramp, from the new surface to the existing surface.

Resumption of Paving Operations – When construction is ready to resume, the following procedure is used to form a suitable transverse joint:

1) The taper of material is removed along with the board or paper.

2) A straightedge is used to check the longitudinal grade of the mat. Because the paver was running out of material as it laid the last few feet of mat, it is possible that those last few feet taper slightly from the specified level of the mat. If this is the case, a new transverse edge must be cut behind the point where the taper begins.

3) The vertical face of the mat is tack-coated.

4) The paver is backed up to the edge of the mat and the screed rested on the mat surface.

5) The screed is heated while it rests on the mat. This provides some heat to the material at the edge of the mat.

6) The heated screed is raised and shims as thick as the difference between the compacted and the uncompacted mat (approximately 25% of the compacted thickness) are positioned under its ends.

7) The truck with the first load of HMA is backed carefully to the hopper. During discharge of the mix from the truck bed to the paver, it is essential that the truck not bump the paver and cause it to move.

8) The paver starts forward in a low gear.

9) Once the paver has moved away, excess HMA is cleaned off the surface of the mat and the smoothness of the joint is checked with a straightedge.

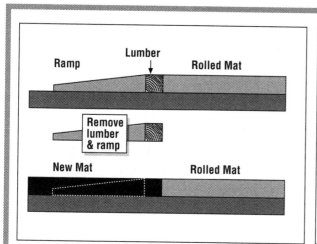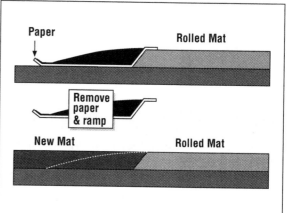

Figure 5.28 Bulkhead Transverse Joint (left) and Papered Transverse Joint (right)

10) If the joint is satisfactory, a 150-mm (6-in.) width of the fresh mix is rolled transversely and the joint checked for smoothness. If the joint is satisfactory, transverse rolling is continued in 150 to 300-mm (6 to 12-in.) increments until the entire width of the roller is on the new HMA. If straight edging shows an uneven joint, the surface of the new mat must be scarified while still warm and workable. Scarification is done with the fine side of the lute. Excess material can then be removed or additional material added, and the joint rolled. During rolling, timbers should be placed along the edges of the mat to prevent the roller from driving off the longitudinal edge and distorting it.

Longitudinal Joints Longitudinal joints occur wherever lanes are placed. There are two types of longitudinal joints: hot joints and cold joints. The vertical seams should be staggered with the surface course joint occurring at the center of a 2-lane pavement (Figure 5.29).

- *Hot Joints* – Hot joints are formed by two pavers operating in echelon. The screed of the rear paver is set to match the grade or thickness of the unrolled edge of the first mat placed. The advantages of a hot joint are that the two mats are automatically matched in thickness, the density on both sides of the joint is uniform because both sides are compacted together, and the hot mats form a solid bond. The disadvantage is that traffic cannot move in one of the lanes while the other is being paved. Both lanes are blocked simultaneously. This process is more applicable to new construction.

- *Cold Joints* – In building a cold joint, one lane is placed and compacted. At a later time, after the HMA in the first lane has cooled, the companion lane is placed against it and compacted. Special precautions must be followed to ensure a joint of good quality.

  – The edge to be joined should be tack coated.
  – The paver screed should be set to overlap the first mat by 25 to 50 mm (1 to 2 in.).
  – The elevation of the screed above the surface of the first mat should be equal to the amount of roll-down expected during compaction of the new mat.
  – For large aggregate mixes, the coarse aggregate in the material overlapping the cold joint should be carefully removed and wasted. This leaves only the finer portion of the mixture to be pressed into the compacted lane at the time the joint is rolled. For mixes with smaller coarse aggregate, such as surface courses, the overlapping material should be pushed with a lute into a hump over the joint area prior to compaction.

Summary of Placing Procedure During construction of a hot mix asphalt pavement, the quality control team has the responsibility to see that contract plans and specifications are met and that the contractor has every opportunity to perform the work in the most cost-effective manner. All paving operations should be preceded by a construction conference, during which details of the work and the test results can be discussed and questions can be answered.

Hot mix asphalt can be placed on several types of surfaces. In all cases, the surface must be properly prepared – free of dust, debris, and excess moisture before paving begins. All existing pavement distress must be corrected prior to overlay construction. Before each paving sequence begins, the agency should assure that the operation is in compliance with the project specifica-

Figure 5.29 Staggering Longitudinal Joints

tions and issue an approval. A check of the grade and alignment of the surface to be paved should be made to be certain there is agreement with the project plans and profile.

Prime coats and tack coats are applied by a calibrated asphalt distributor at a rate determined by the flow of asphalt from the tank and the distributor's forward speed. Proper spraying temperatures and application rates should be used in all prime and tack coating operations.

The paving machine consists of two major units: a tractor unit and a screed unit. The tractor unit includes the power plant and all control systems for delivering power to systems throughout the paver. The screed unit places the HMA mat and includes controls for regulating mat thickness.

The personnel charged with the responsibility of obtaining a satisfactory product should be familiar with the various types of pavers, screeds, screed controls, and grade control systems. They should also be aware of the importance of uniformity and balance in all paving operations.

Load tickets collected during placement serve as part of the final records for the project. All diary notes, sketches, and calculations for compliance should be retained.

The screed person should check the temperature often to assure uniform quality of mat texture, density, and smoothness. He should make sure that all joints are smooth and tight and that all handwork blends in with the over-all texture of the project.

Fume Recovery System

In July 1996, NIOSH held a public meeting to discuss the use of engineering controls for reduced worker exposures to asphalt fumes during paving operations. On the basis of promising preliminary research, representatives of industry and labor wanted to move forward with field implementation of engineering controls. As a result of this work, industry designed and constructed a fume collector system that would remove 80% of the indoor capture efficiency of fumes from industrial pavers. Industrial pavers are those weighing over 7,258 kg (16,000 lbs.).

The fume recovery system is deigned to collect fumes given off by HMA and discharge them from the operator work area (See Figure 5.30). The auger access hood sections can be removed to gain access to the main spreading augers. This may be necessary for checking and maintaining the augers. The system is designed to meet the minimum standards with or without the auger hood sections installed; however, to operate at maximum effectiveness, they should be installed during operation.

The fume recovery fan starts automatically when the engine is started and continues to run while the engine runs. The vacuum gauge should be checked each day before paving begins. With the engine at full throttle, the gauge should between 25 to 50 mm of mercury. A reading below this indicates a loss of vacuum and that the system is not working properly.

The paving equipment manufacturer can be consulted about the availability of a kit for existing pavers.

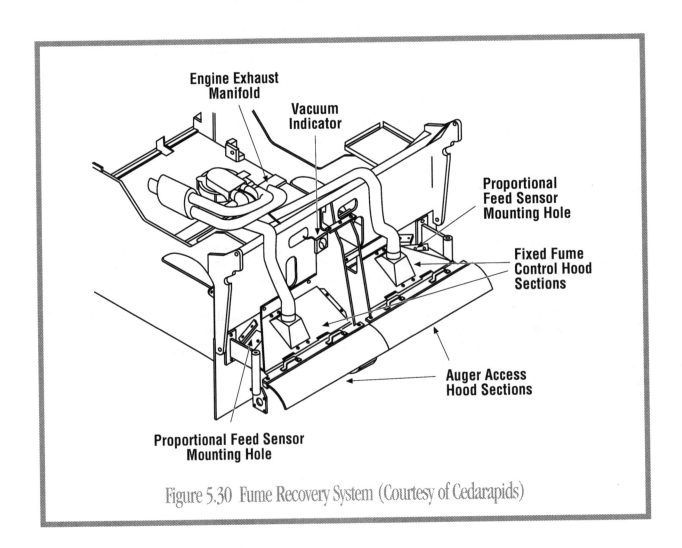

Figure 5.30 Fume Recovery System (Courtesy of Cedarapids)

CHAPTER 6

COMPACTION

Introduction

The last aspect of the hot mix asphalt (HMA) construction process is compaction. Compaction is the process by which freshly placed HMA is compressed to reduce the in-place air voids and produce a smooth, long lasting pavement. This is a very crucial step in the construction process. Proper and effective compaction is necessary in order to prevent the excessive intrusion of air and water into the pavement structure. Research has shown that when in-place air voids are in excess of 8 percent, the voids tend to be interconnected throughout the pavement. Circulation of air and water (both of which contain oxygen) through a poorly compacted pavement can greatly reduce its longevity and, in extreme cases, cause catastrophic failure (see Figure 6.01). The ability of air and water to move freely through the pavement accelerates the oxidation process and causes "aging," or hardening. This aging leads to more brittle asphalt binders, therefore reducing durability in the pavement. A porous mixture that allows excessive moisture to penetrate promotes rapid aging and raveling of the mixture. If moisture is trapped in the pavement it could lead to stripping. This intrusion of air and water can be greatly reduced—if not eliminated—when the pavement is properly compacted to an in-place air void level of 8 percent or less. Air voids are less likely to be interconnected when pavements are compacted to less than 8 percent air voids.

Some pavements, such as Open Graded Friction Courses, are designed with in-place air void contents as high as 20 percent. These types of mixtures generally consist of a very high quality, coarse aggregate structure coated with a thick asphalt film. This thick asphalt film resists the effects of water and air in the mixture, allowing it to perform under harsh circumstances. These types of mixtures require adjustment in the rolling procedures. Normally, only two or three passes of a static roller are required to "set" the mixture. However, when compacting extremely coarse mixtures, other adjustments must be made and will be discussed in following sections.

Compaction Principles

▶▶ General Mechanics of Compaction

The mechanics of compaction involve three main forces at work. They are: the compressive force of the rollers, the resistive forces within the mixture, and the supporting forces exerted by the stable surface beneath the mat, be it subgrade, aggregate base or pavement. Compaction of the HMA will be completed when these three forces reach equilibrium.

Construction of a quality roadway begins from the bottom up. The subgrade (or other stable surface beneath the mat), along with the internal friction of the mixture, resists the downward compaction force of the rollers. The underlying layers must be able to support the forces exerted by the roller in order to overcome the internal resistance of the mixture. If the subgrade is not firm and stable, the HMA will not be confined and compaction will not be achieved (see Figure 6.02). This also applies to the HMA; if it is not stable enough to resist the compaction forces, it will tend to displace and not com-

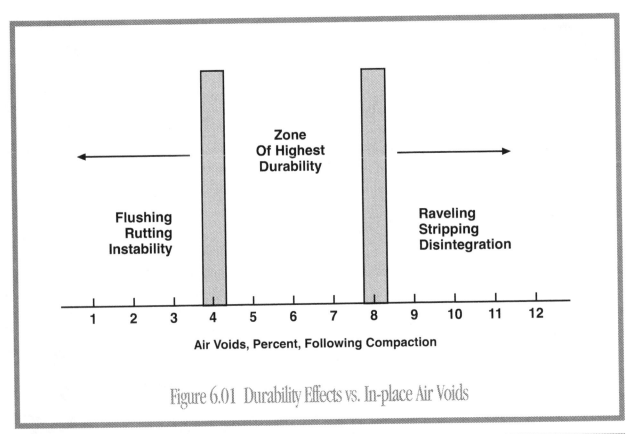

Figure 6.01 Durability Effects vs. In-place Air Voids

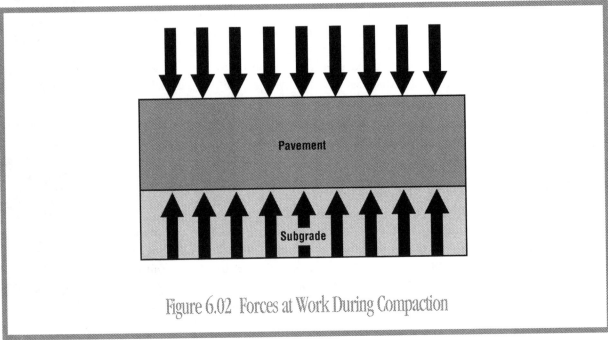

Figure 6.02 Forces at Work During Compaction

pact to the desired air void level. Finally, if the rollers do not exert enough force to overcome resistance within the mixture, the pavement will not be sufficiently compacted. As current and future traffic loading requires harsher mixtures, more compaction effort is needed to overcome the internal resistance of the mixture to reduce the air voids to acceptable levels. Awareness of these reactions is necessary to establish a proper compaction process.

➤➤ HMA Factors Affecting Compaction

Internal resistance of the HMA is dependent on:

- Mixture properties and characteristics.
- Environmental conditions.
- Layer (lift) thickness.
- Subgrade and bases.

Each of these factors will be discussed in detail in the following sections.

Mixture Properties and Characteristics Resistance to compaction of the HMA is important in determining the amount of compactive effort that is required to reduce the air voids in the pavement to an acceptable level. There are many variables that determine this level of resistance, including:

- Aggregate properties.
- Asphalt binder properties.
- Temperature.

Aggregate Properties – It is important to remember that HMA is a combination of aggregate and asphalt binder. The aggregate acts as the structural skeleton of the pavement and the asphalt binder as the glue of the mixture. The aggregate used in the mixture greatly affects how the HMA reacts on the roadway. Gradation, surface texture and angularity are the primary characteristics that affect workability of the mixture. As the maximum aggregate size increases or the amount of coarse aggregate increases, the mixture becomes more resistant to compaction. This decreases the workability of the mixture and requires more compactive effort in order to reduce the in-place air voids. An increase in surface texture and angularity also creates the same effect. A crushed, rough surfaced, cubical aggregate provides more particle-to-particle friction than round, smooth, natural aggregate. An excessive amount of natural sands around the 0.60 mm (No. 30) sieve size can yield especially tender mixtures. Crushed aggregates are generally much more stable than natural aggregates; however, when some aggregates are crushed, they fracture into very smooth or flat and elongated particles. These types of aggregates will not provide as much interparticle friction.

The dust content, or material passing the 0.075 mm (No. 200) sieve, will also affect the compaction process. It is the combination of fines and asphalt that provides the binding mastic in HMA pavements. The mixture should contain sufficient fines to combine with the asphalt to produce the necessary cohesion when the mixture cools. The addition of a mineral filler, such as lime, will help to counter tenderness or "slow setting" properties of a mixture that contains too much natural sand. However, if a mixture contains too many fines it may become "gummy" and unworkable.

Asphalt Binder Properties – Asphalt is a thermoplastic, temperature susceptible material. As the temperature of asphalt increases, the viscosity decreases, that is, it gets thinner. As it cools, it gets thicker or more viscous. When the asphalt binder is fluid (hot), it easily mixes and coats the aggregate particles. It also acts as a lubricant that facilitates compaction of the mixture on the roadway. As the asphalt binder cools, it becomes stiffer and binds the aggregates to provide a durable long lasting mixture. For mixtures with neat asphalts, once the temperature of the mixture cools to around 85°C (185°F), the asphalt binder thickens rapidly. Compaction of HMA should be complete before the mixture temperature falls below 85°C (185°F). Continuing to compact can damage the pavement and actually increase the air void level in the mixture.

Superpave specifications may require that premium asphalt binders be used in many areas. These premium binders may require modifiers to achieve the performance desired. Some modified binders have elevated softening points that may require the HMA to be compacted well above 93°C (200°F). It is difficult to predict what the minimum temperature will be. Experience, field testing, and information from the asphalt supplier will provide the best source of knowledge to attain a quality pavement.

Temperature – The temperature of a mixture is perhaps the most important property in obtaining density, since the viscosity of an asphalt binder is controlled by its temperature. The temperature at which HMA is produced is the first controlling factor that will determine the compaction temperature on the roadway. During the initial mixture design, mixing and compaction temperatures are determined from curves on the temperature-viscosity chart. The laboratory mixing temperature is the temperature at which the asphalt binder reaches a viscosity of 0.17 (±0.02 Pa-s 170 ± 20 centi-stokes). The laboratory compaction temperature is that at which the asphalt binder reaches a viscosity of 0.28 ±0.03 Pa-s (280 ± 30 centistokes). These viscosity ranges, however, are not necessarily valid for modified asphalts. The mixing and compacting temperatures obtained are intended for laboratory mix design procedures. Even so, a good starting point for the plant mixture production temperature is the laboratory mixing temperature.

It is not realistic, however, to expect the pavement compaction process to be completed at the laboratory compaction temperature. The laboratory compaction temperature can often approach or exceed 150°C (300°F). Typically, the difference between laboratory mixing and laboratory compaction temperatures is relatively small, in the range of 10°C (18°F). This would not allow sufficient time to load, haul, place and compact the asphalt mixture. However, construction specifications are usually written to require the compaction process to be completed before the in-place temperature of the mixture cools to 85°C (185°F). This should allow enough time for the pavement compaction to be completed.

The same mixing and compaction temperatures may not apply to modified binders in either laboratory mix design or production and compaction. Modified binders may not reach the mixing viscosity of 0.17 ±0.02 Pa-s (170 ± 20 centistokes) until heated to 150°C (350°F) or higher. In addition, modified binders may become stiff well before their temperature drops to 85°C (185°F). When working with these special binders, consultation with the binder manufacturer is essential in determining construction parameters.

In general, mixing temperatures should be kept as low as possible in order to obtain uniform mixing and coating of the aggregate and asphalt, yet high enough to provide field crews sufficient time to place and compact the HMA on the roadway. High performance pavements, with coarse aggregate structures and modified binders, may require that compaction be completed at temperatures well above 93°C (200°F). Close coordination between plant production, temperature, length of haul, time for placement, and compactive effort will be necessary to achieve a properly constructed pavement.

Environmental Conditions The construction of quality pavement structures is highly dependent on the conditions under which the pavement is placed. Ambient air temperature, wind, humidity and the temperature of the surface upon which the HMA is being placed can seriously affect the cooling rate of the mixture. The placement and compaction of HMA is basically a race against time (see Figure 6.03). Cool air temperatures, high humidity, strong winds and cool surfaces can shorten the time in which compaction must take place. Increasing plant mix temperature, covering hauling units, minimizing haul length and shortening windrows in front of pick up machines can all minimize the effects of the environment.

Layer Thickness All asphalt mixtures cool with time. The greater the surface area of the mixture, the faster the environment can cool the mixture. Thick layers, or lifts, have less material exposed to the air and subsurface in relation to their volume, and therefore cool slower. Generally, it is easier to achieve required density in thicker lifts of HMA than in thinner ones. This is because the thicker the mat, the longer it retains heat, thus increasing the time during which compaction can take place. This principle can be used to advantage when rolling lifts of highly stable mixtures that are difficult to compact, or when paving in weather that can cause rapid cooling of thin mats. The most effective way to slow the rate of cooling is to keep the mixture in as large a mass as possible. Thicker layers can permit mixtures to be placed at lower temperatures because of the reduced rate of cooling.

Subgrade and Bases The subgrade or base must be firm and non-yielding under the haul trucks and other construction equipment. Subgrades or bases that show movement under trucks or construction equipment will need additional compaction work or some type of remedial work to overcome the softness. The remedial work could be lime or portland cement stabilization, or in certain circumstances, removal and replacement with a more suitable material. A yielding subgrade or base would require a thicker HMA pavement in order to support the traffic loading. Haul trucks may also be limited in size and weight to prevent pumping action of basement materials.

➤➤ *Cold Weather Paving*

As previously discussed, maintaining heat in the mix is critical in achieving density and constructing a quality asphalt pavement. In many areas however, it is necessary to pave when conditions are not exactly favorable. Many times in the early spring or late fall and winter, temperatures can severely hamper or halt the construction process. When paving in cold weather conditions, construction procedures usually need to be altered in order to achieve the specified density. A few things that can be done to achieve density are:

- Work the breakdown and intermediate rollers as close to the paver as possible.
- Cover loads during hauling (warm and cold weather paving).
- Insulate the truck beds.
- Increase the plant temperature.
- Decrease plant production rate.
- Minimize or eliminate windrows and pickup machines.
- Reduce paving speed.
- Increase the number of rollers.
- Increase the layer thickness.

The basic strategy is to maintain as much heat as possible in the mixture long enough to facilitate attainment of the required density. However, even by observing all of the above recommendations and achieving the specified level of compaction, detrimental effects can occur when paving during cold weather. As the HMA cools, it cools at different rates. The exposed surface and the surface contacting the cold subsurface will cool at a much faster rate. The top and bottom of the layer will rapidly cool to the surrounding temperature, while the middle portion of the layer (which is often where the temperature is measured) remains much warmer. It is theorized that, when the lift is compacted, the top surface can develop micro cracks as the mat compresses. These micro cracks allow moisture to enter the pavement surface, and moisture

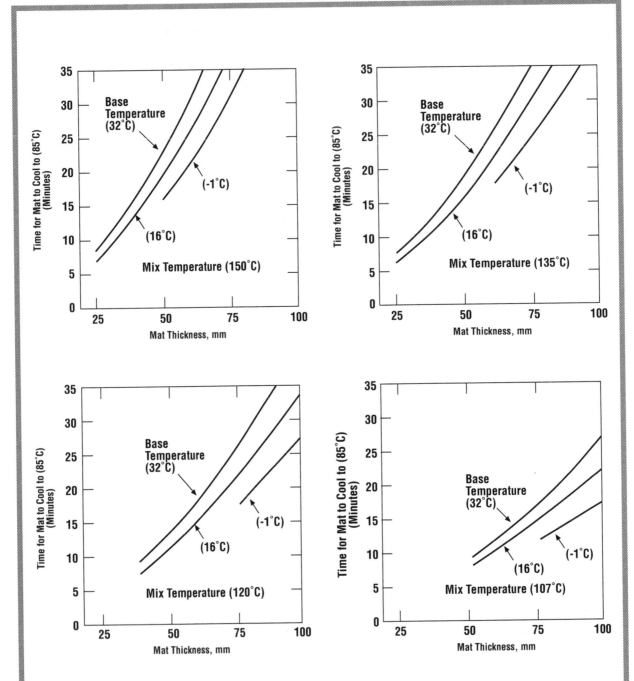

Wind velocity – 10 knots. Atmospheric temperature – same as base.

Note: "Base Temperature" is the temperature of the surface upon which the asphalt mat is placed.

85°C is the temperature of the mat measured 6 to 12 mm below the mat surface. The average temperature of the entire mat thickness when this temperature is reached, is approximately 80°C.

Placing thicknesses less than those shown by the curves is not recommended on sugrades of -1°C (base temperature).

(Conversion: 25.4 mm = 1 in, °F = [(9/5)°C]+32)

Figure 6.03 Time Allowed for Compaction

damage can be expected. As a result, surface raveling could occur. Raveling is the process where surface fines and binder are lost and "flake" away. The resulting surface will appear porous and rocky as the large aggregate remains in place. This phenomenon is more severe when using marginal or "poor" aggregates and when paving in the fall. Early season paving problems tend to be less prevalent since traffic during the summer seems to "knead" the surface back together.

Perhaps the most common consequence of cold weather paving is inadequate densification of the pavement layer, especially thin surface courses. The lack of adequate density in late paving season often results in raveling, and sometimes disintegration, of the surface course within a few months after paving.

Equipment

➤➤Rollers

Self-propelled rollers are required for the compaction of HMA. Towed-type rollers should not be used. Hand-held or vibrating plate compactors can be used in small, inaccessible areas. Typical self-propelled compaction rollers consist of the following types:

- Static Steel-wheeled
- Pneumatic-tired
- Vibratory Steel-wheeled

Static Steel-Wheeled Tandem Rollers Static Steel-wheeled rollers (Figure 6.04) have steel drums generally mounted on two tandem axles. Typical tandem static steel-wheeled rollers vary in weight from 2.7 to 12.7 metric tons (3 to 14 tons). On many types of static rollers, adding or removing ballast can vary the weight. For streets, highways and other heavy-duty pavement construction, a minimum gross weight of 9 metric tons (10 tons) is required. Static steel-wheeled rollers can be used for breakdown (initial) rolling, intermediate rolling, or finish rolling.

Static steel-wheeled tandem rollers should provide a minimum of 4.46 kg/mm (250 lb/in) mass (weight) per width on the compaction roll (drive wheel) when used for breakdown or intermediate rolling. Steel drums that are grooved, pitted, worn or warped should not be used. Steel drums should be checked with a sharp metal straightedge. Surface imperfections or rust may tend to pick up hot mix as it is being rolled. Water spray bars and wetting pads are used to prevent adhesion to the drums. Scraper bars are also utilized to remove any particulate matter that may build up on the drum.

Figure 6.05 illustrates the force exerted by a steel-wheeled roller on an HMA mixture when the surface under the mixture is firm. The arrows indicate the direction of lines of force through the mat. Notice that the lines of force directly under the roller extend through the pavement to the subgrade. The firm subgrade exerts a resistive force upward. The mixture between the roller and subgrade is compacted as a result of the two forces acting in opposite directions.

Static steel-wheeled rollers typically have one powered drum, often referred to as the drive wheel, and one non-powered steering drum called the tiller wheel. Orientation of the roller is critical in most cases, particularly during initial compaction. The roller's direction of travel should be such that the powered wheel passes over the uncompacted mixture first.

Figure 6.04 Static Steel-wheeled Roller (Courtesy of Ingersoll-Rand)

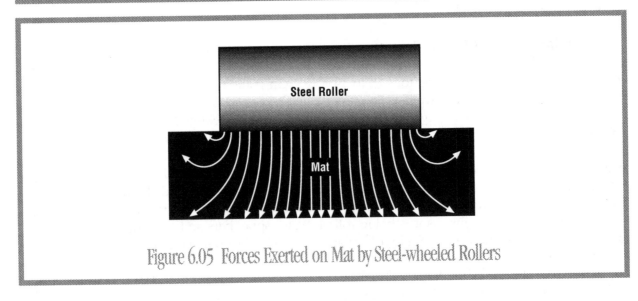

Steel Roller

Mat

Figure 6.05 Forces Exerted on Mat by Steel-wheeled Rollers

Figure 6.06 illustrates the correct use of a steel-wheeled roller. The drive wheel is ahead of the tiller wheel in the direction of travel on the uncompacted mixture. It can be seen that there is a vertical force downward caused by the weight of the wheel. The arrows concentric with the drive wheel represent the rotational force on the wheel, which is transmitted to the mixture as the roller is propelled. This concentric force tends to move the mixture under the wheel rather than to push it away. The resultant of these forces approaches a direct vertical force from the drive wheel.

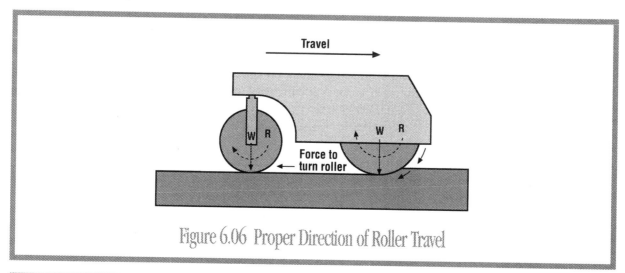

Figure 6.06 Proper Direction of Roller Travel

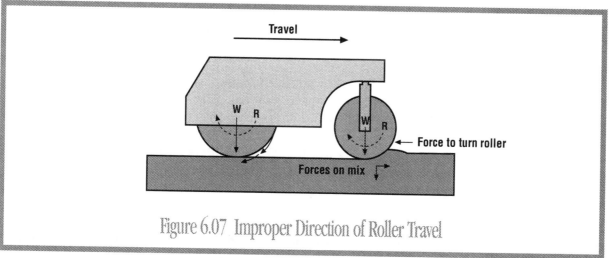

Figure 6.07 Improper Direction of Roller Travel

Figure 6.07 illustrates a steel-wheeled tandem roller being used incorrectly on an HMA mixture. The tiller wheel is in front, in the direction of travel. This can be a critical mistake on some mixtures, particularly during the breakdown pass. Since the tiller wheel is a dead wheel without power of its own, there is a tendency for it to push the mixture away from itself, causing a wave in front. An analysis within the mixture reveals two forces. One is a vertical force downward, and the other is a horizontal force forward. For compaction of a mixture, the desirable movement of all aggregate particles is vertically downward. Little, if any, densification occurs as a result of the horizontal movement within the mixture. Horizontal movement of the mixture can actually result in a reduction of density. Excessive horizontal movement can cause hairline cracks to appear in the surface of the mat. Differential cooling throughout the lift can make this problem worse, causing what is often called "heat checking."

There are other reasons for the drive wheel to travel on the mixture ahead of the tiller wheel, particularly during the breakdown pass of the roller. Since the drive wheel has the largest diameter, it presses with a flatter contact surface on the mixture. Therefore, the horizontal force from the wheel is minimized. Because the drive wheel has a larger diameter than the tiller wheel, it does not sink as deep into the mixture. This also reduces the horizontal component of force imparted by the wheel. The drive wheel is the heaviest wheel and is considered to be the compaction wheel. Since the best time to compact is when the resistance is the least, while the mixture is hot, the breakdown pass should be done with the compaction wheel on the mixture first.

When rolling on steep grades, the above mentioned procedure may need to be altered. On steep grades, the tiller wheel may need to be kept in front, following the direction of paving. The lighter tiller wheel provides initial compaction to increase the stability of the mixture before the heavier drive wheel contacts the mixture. This results in a reduced tendency for the mixture to move downhill.

The weight of the roller is transmitted to the mixture through the contact pressure that is exerted under the drums. Therefore, the contact pressure under the drums should not exceed the supporting capability of the mixture being compacted. Usually, heavier rollers can be used on harsher, more stable mixtures, particularly for breakdown passes. Somewhat lighter rollers may be necessary on less stable mixtures.

Pneumatic-Tired Rollers Pneumatic-tired rollers have rubber tires instead of steel tires or drums. They generally feature two tandem axles, with 3 to 4 tires on the front axle and 4 to 5 tires on the rear axle (Figure 6.08). The wheels oscillate; that is, they move up and down independently of one another.

Pneumatic-tired rollers can be ballasted to adjust their gross weight and, depending on size and type, may vary from 9 to 32 metric tons (10 to 35 tons). More important than gross weight, however, is the weight per wheel, which should range from 1360 to 1590 kg (3000 to 3500 lbs.) when the pneumatic-tired roller is to be used for breakdown or compaction rolling.

Pneumatic-tired rollers may be equipped with 380, 430, 510, or 610-mm (15, 17, 20, or 24-in.) wheels, and should have smooth tires for asphalt compaction. The tires must be inflated to equal pressures, with variation not exceeding 35 kilopascals (5 psi), to apply uniform pressure during rolling.

Pneumatic-tired rollers may be used for breakdown and intermediate rolling for compaction, and for conditioning a finished asphalt surface. Rubber-tired rollers have traditionally been utilized for intermediate rolling. With increased traffic and coarse, dense-graded mixtures, the pneumatic type rollers may provide the additional compactive effort needed to achieve specified density. Pneumatic type rollers are not recommended for breakdown rolling on steep grades.

Rolling for compaction and surface conditioning are two different processes and require different operating procedures.

Figure 6.09 illustrates the action of a pneumatic-tired roller when used for breakdown and intermediate rolling. The arrows illustrate typical lines of force in the mat.

As in the case of steel-wheeled rollers, when pneumatic-tired rollers are used, the mixture being compacted must be adequately confined for proper densification. Uniform subgrade strength may be more critical when pneumatic rollers are used because the individual wheels can exert high stress on small areas of subgrade weakness that wide, rigid steel drums tend to bridge.

When a rubber-tired roller is used for breakdown rolling, very little horizontal movement of the mixture occurs in the direction of travel. This is because each tire flattens slightly as it drives over the mixture permitting almost all of the compactive force to be exerted vertically on the mat. Horizontal movement of the mixture in the direction of travel occurs only if the tire diameter is too small and the tire sinks into the mixture. Excessive movement is an indication that the HMA is unstable or the roller being used is unsuited for breakdown rolling. There is some horizontal movement of the mixture under a pneumatic tire, but it tends to be at right angles to the direction of travel. It may cause small bumps in the mixture immediately adjacent to the tire.

Figure 6.08 Pneumatic (Rubber)-Tired Roller (Courtesy of Dynapak)

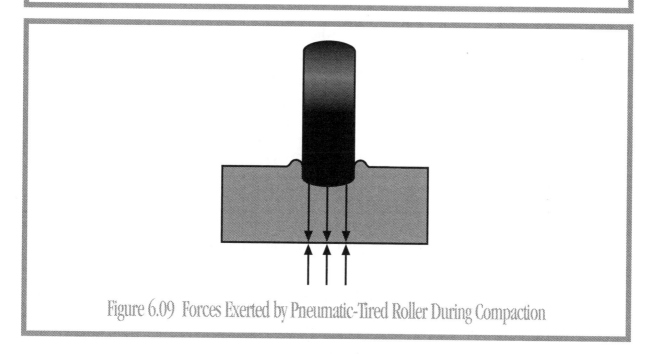

Figure 6.09 Forces Exerted by Pneumatic-Tired Roller During Compaction

These small bumps normally are of no significance and will be rolled out by subsequent passes. Reducing the tire pressure will reduce this lateral displacement. Additional passes should eliminate such bumps as well as any tire marks (ruts) in the mat surface. The surface may still look irregular but this appearance is mostly cosmetic. Because the tires must be allowed to heat up to avoid mixture sticking to them during breakdown and intermediate rolling, water is typically not used on the tires of a pneumatic roller. Mixture will stick to the tires during the warm-up period, but once they are hot, this will cease. Skirts placed around the tires will shorten the warm-up period and help keep the tires hot, particularly in cool or windy weather. As with steel-wheeled rollers, the drive wheels of the pneumatic-tired roller should be toward the paver.

Desirable pneumatic-tired roller requirements for breakdown and intermediate compaction are:

- A weight per wheel of 1360 to 1590 kg (3000 to 3500 lbs.).
- 510 mm (20-in.) minimum wheel diameter.
- Tire inflation pressure of 483 to 517 kPa (70 to 75 psi) when cold, and 620 kPa (90 psi) when hot.

These recommended tire pressures are applicable for most mixtures but can be reduced if necessary for mixtures with low stability.

The kneading action of a pneumatic-tired roller can also be employed to improve or toughen an asphalt pavement surface after normal paving operations have been completed. The rolling can be performed as much as two weeks after the pavement has been placed, provided the weather is warm and the pavement surface temperature is at least 38°C (100°F). The kneading operation can reduce pavement permeability and can increase pavement resistance to scuffing or abrasion by traffic.

Pneumatic-tired rollers are also ideal for correcting heat checking in the mat surface. Heat checking is the appearance of short, 50 to 100-mm (2 to 4-in.) long, disconnected hairline cracks after one or more passes of the static steel-wheeled roller.

When a pneumatic-tired roller is used for kneading a finished asphalt surface, the desirable requirements are:

- A 680 kg (1500-lb.) minimum weight per wheel.
- A 380 mm (15-in.) minimum wheel diameter.
- Tire inflation pressure of 345 to 415 kPa (50 to 60 psi).

The ability of pneumatic-tired rollers to provide a tighter, more traffic-resistant surface than steel-wheeled rollers was recognized years ago when they were first employed for the intermediate rolling of HMA pavements. Subsequent testing indicates that pneumatic-tired rollers will achieve about the same pavement density as steel-wheeled rollers.

In HMA overlay construction, the first (leveling) course is very often placed on an irregular surface that has been rutted or disfigured by traffic. The ability of the pneumatic-tired roller to apply uniform pressure over its entire width makes it desirable to be used for this course because it will apply compactive effort where it is needed most—in the wheel tracks. The bridging action of steel-wheeled rollers may prevent them from being as effective in similar situations.

Vibratory Rollers Vibratory rollers provide compactive force by a combination of weight and vibration of their steel compaction drums. Those used for compaction of HMA are self-propelled and vary in weight from 6 to 15 metric tons (7 to 17 tons). There are two basic models: the single drum units (Figure 6.10), and the double drum (tandem) units (Figure 6.11).

Either steel or pneumatic-tired wheels can provide propulsion for single drum vibratory models. Both drums usually provide propulsion for the double drum models. The drums on vibratory rollers vary from 0.9 to 1.5 meters (3 to 5 ft.) in diameter and 1.2 to 2.4 meters (4 to 8 ft.) in width. Their static weight in terms of drum width is generally from 2.9 to 3.2 kilograms per millimeter (160 to 180 lb. per in.).

Figure 6.10 Single Steel-Wheeled Vibratory Roller (Courtesy of Bomag)

The engine providing power for propulsion also powers the hydraulically driven vibrating unit. Vibrations are generated by a rotating eccentric weight inside the drum, the speed of which determines the frequency, or vibrations per minute (vpm), of the drum. The weight and distance from the shaft of the eccentric determines the amplitude, or amount of movement, of the impact force that is generated. Both the frequency and amplitude of the vibrations are controlled independently of roller travel and engine speed.

The vibration frequency of rollers used for HMA compaction is generally between 2000 and 3000 vibrations per minute, depending on the model and manufacturer. Some models provide only one or two specific frequency settings, while others may provide a full range of frequencies within certain limits, such as 2200 to 2800 vpm.

Vibratory rollers achieve compaction through a combination of three factors:

- Weight
- Impact forces (roller vibration)
- Vibration response in the mixture

Weight – Weight has been discussed in connection with steel-wheeled tandem rollers and pneumatic-tired rollers. The impact forces are those generated by vibration of the compaction drum and are regulated by controlling the frequency and amplitude of the vibration. The impact

Figure 6.11 Double Drum Vibratory Roller (Courtesy of Caterpillar)

Vibration Amplitude

Higher

Lower

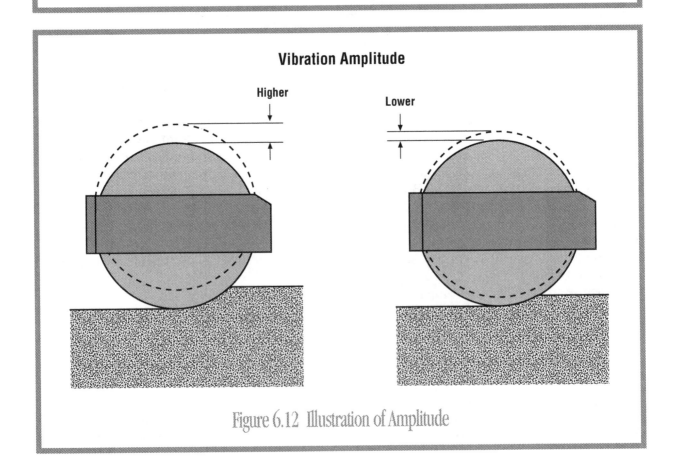

Figure 6.12 Illustration of Amplitude

force required to obtain specified densification of the mat varies with the temperature and properties of the mixture, the thickness of the mat and the support provided by the surface. It will also vary with the drum diameter and width and the roller's static weight and dynamic (impact) force.

The vibration response in the mixture is the reaction to the forces exerted upon it. As with other types of rollers, the mixture will compact easily or with difficulty depending on its temperature, cohesion, particle shape and texture, confinement and other factors. The thing that is different when vibratory rollers are used is that repetitive dynamic forces are being exerted on the mixture.

To use a vibratory roller effectively, it is necessary to have some understanding of the movements which influence the compaction forces, that is, frequency and amplitude of the vibrations.

Frequency – Roller drum vibrations are produced by off-center weights, called eccentrics, on a spinning shaft.

The speed of the shaft sets the frequency (the number of vibrations or downward impacts per minute).

Frequency is defined as the number of cycles per minute – a single cycle being one full turn of the eccentric. The eccentric is an off-center weight fastened to a shaft, typically inside the drum. As the shaft spins, the eccentric creates an outward force. The heavier the eccentric the greater the force produced. The farther it is from the shaft the greater the force produced. And the faster it spins, the greater the force produced.

As the roller moves ahead, its vibrating drum produces a rapid sequence of impacts on the surface. These impacts are equal to the frequency of vibration. For any given roller speed, the higher the frequency used, the closer the impact spacing will be and the smoother the surface will be. The manufacturer's advice on frequency should be used for each roller.

Amplitude – The roller drum moves up and down as it vibrates (Figure 6.12). When it changes direction in its up-and-down movement, it is momentarily at rest, just as the roller itself is at rest when it changes direction. Amplitude, then, is the greatest movement in one direction of a vibrating roller drum from a position at rest. The weight and distance of the eccentric from the shaft and the weight of the roller drum control the impact force. For any given drum weight, the heavier the eccentric is, and the farther away it is from the shaft, the higher the amplitude will be. On most heavy tandem vibratory rollers, the amplitude can be varied by the operator to suit paving conditions. For each roller, the manufacturer's advice on amplitude should be used.

Use of Vibratory Rollers

To ensure smoothness under vibratory compaction, the frequency and roller speed should be matched so that there will be at least ten downward impacts of the vibration per foot of travel of the roller. The relationship between speed and frequency is illustrated in Figure 6.13. As the speed of the roller increases for a given frequency of vibration, the spacing of the impacts grows farther apart. In asphalt mixtures, it is generally agreed that the most desirable method is to use the maximum rated frequency with the speed of the roller adjusted to provide the desired impact spacing.

Figure 6.14 illustrates four different modes of using a vibratory roller equipped with two vibrating drums. The first mode shows the roller being used without vibration. It acts as a static steel-wheeled tandem roller. This is the mode utilized when using a vibratory roller for finish rolling. The second mode shows the use of vibration on the trailing drum with the leading drum in the static mode. This mode may be desirable on mixtures that have lower stability. The third mode illustrates the use of vibration with both drums, which is used on a stable mixture in order to achieve the maximum compactive energy. The fourth mode illustrates vibration only on

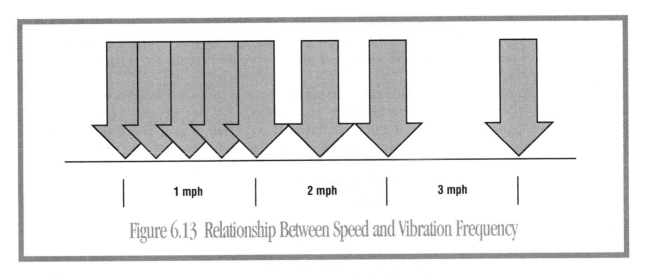

Figure 6.13 Relationship Between Speed and Vibration Frequency

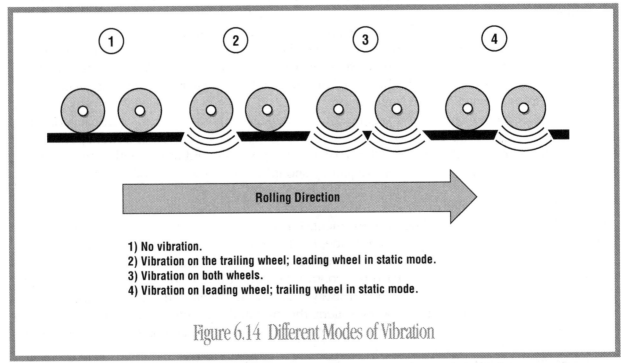

Rolling Direction

1) No vibration.
2) Vibration on the trailing wheel; leading wheel in static mode.
3) Vibration on both wheels.
4) Vibration on leading wheel; trailing wheel in static mode.

Figure 6.14 Different Modes of Vibration

the leading drum of the roller. This mode is used to achieve compaction with the leading drum while the trailing drum in the static condition provides a smoother finish. The selection of the mode of operation should be tailored to the mixture and the conditions of the project.

The energy imparted by the vibratory compactor is absorbed in the mixture being compacted. Controlling the amplitude permits the operator to vary the vibratory force, and therefore, to vary the energy imparted to the mixture. An amplitude adjustment may be necessary for each modification in the mixture being placed. For example, a change in the layer thickness, mixture temperature, mixture gradation, filler content or asphalt content may require adjustment in both the amplitude and frequency being used. It is important that the roller should be vibrating only when it is moving. If vibration continues while the roller is standing still or changing direction, each vibrating drum will leave an indentation in the pavement at the stopping point. Most modern rollers have automatic vibration cut-off devices that actuate when the roller stops moving.

Table 6.01 Guide for Setting Vibratory Compactor Controls In Relation to Layer Thickness

Layer Thickness	Frequency	Amplitude
Thin	Maximum	Low
Thick	Maximum	High

Normally, vibration is not used in the compaction of very thin lifts. This is particularly true with sandy or tender type mixtures. In very thin lifts, there is insufficient material to absorb the energy imparted by the vibrating rollers. The energy passes through the mixture being compacted and rebounds from the surface of the pavement below. It re-enters the mixture and de-compacts the mat. For situations of this type, the vibratory roller should be used in the static mode. Table 6.01 shows guidelines for compacting thinner lifts that are substantial enough to hold a vibratory roller and for compacting thicker lifts.

Special attention may be necessary when using a vibratory roller on steep grades. For tender or low stability mixtures, the initial breakdown passes may need to be run in the static mode to prevent displacement of the mixture. Once initial stability is established, vibration can begin.

Caution should also be exercised when vibratory rolling is performed on extremely coarse mixtures. The aggregate structure (stone-on-stone) may be fractured under intense rolling. If specified density cannot be obtained with static steel-wheeled and/or pneumatic-tired rollers, vibratory rollers can be used, but the amplitude should be kept at the minimum necessary to achieve the specified density.

Rolling Procedures

The degree of density achieved in HMA is dependent on the amount of compactive effort applied before the mixture cools to 85°C (185°F). The variables that will affect the length of time in which compaction must be accomplished have been discussed. One that has been only briefly mentioned is the rate of mixture production. Increasing roller speed will not compensate for increased production rates; it will simply reduce the amount of compactive effort that is applied to a given area of pavement surface in a given time interval. The rolling speed, whether being used in the vibratory or static mode, should not exceed 5 to 8 kilometers per hour (3 to 5 mph). This rolling speed is the maximum recommended for static steel-wheeled, pneumatic-tired, and vibratory rollers.

Additional rollers will be needed when there is an increase in production if the current rollers cannot achieve the desired compaction on the increased amount of material. The number of rollers provided must be tailored to the conditions of the specific job and be adequate to obtain the desired compaction.

➤➤ Determining Roller Requirements

The exact number of coverages (passes) of a roller or rollers that will be required to obtain adequate density is initially unknown due to uncertainty of the cooling rate of the mat. These

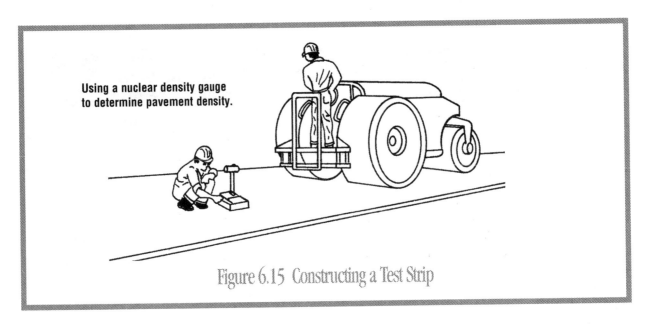

Using a nuclear density gauge
to determine pavement density.

Figure 6.15 Constructing a Test Strip

uncertainties are resolved by the testing that is performed in the initial test strip, or control strip, and by observation during the early stages of the paving operation.

A number of studies have been made on the cooling rates of mixtures under varying conditions of mixture temperature, lift thickness and base temperature. The temperature of the mix and the thickness of the mat can be used to make a fairly accurate estimate of the time interval in which density must be achieved before the temperature of the mix falls below 85°C (185°F) (see Figure 6.03). This estimate can be used to determine the number of rollers required on the job.

A test strip will establish the rolling pattern to assure achievement of the required density and the proper riding quality, and to attain the optimum production rates with the given roller. In most cases where test strips are properly used, the rollers will compact the mix in the remaining pavement to meet the density requirements and produce a good riding surface (see Figure 6.15). Information on how to build a test strip is in the section on Control Strip Density in Chapter 7.

A rolling pattern that will provide the most uniform coverage of the paving lane should be planned and used. Since rollers are produced in a number of widths, a uniform pattern that applies to all rollers and conditions is not possible. For this reason, the best rolling pattern for each roller being used should be determined on a test strip. The following procedure can be used as a guide.

1) Before rolling the test pattern, decide how the roller will be operated for:
   a. Rolling speed.
   b. Lap pattern for paving width.
   c. Number of passes.
   d. Selection of roller operating zone behind the paver.

2) If the test pattern does not pass, a series of new test patterns should be run. The following steps are recommended:
   a. Slow the roller down.
   b. Take a 15-second test with a nuclear density gauge after each pass (or round trip) until maximum achievable density is indicated by the test results.

c. The correct rolling speed is always a balance between rolling fast for productivity and rolling to meet density and finish requirements. Therefore, if the selected speed obtains the required density, but leaves surface blemishes, reduce the speed until blemishes disappear.

d. Roller speeds should not exceed 8 km/h (5 mph).

3) The rolling pattern used on the test strip should be the same pattern that will be used on the remainder of the job.

a. The roller should not be operated slower than it will be operated on the remainder of the job.

b. If the number of roller passes required to obtain density on the test strip is high, the rollers may not be able to keep up with the paver at its normal speed on the rest of the job. In this case, the number of rollers must be increased or the rate of production must be decreased.

4) It is very important to recognize that during the rolling process all operating techniques are governed by mixture behavior. This will vary from job to job and from lift to lift.

➤➤ Sequence of Rolling Operations

As mentioned before, there are three phases of rolling operations. They are:

- Breakdown (initial) rolling – The first pass of the roller on the freshly placed mat
- Intermediate rolling – All subsequent passes by the roller(s) to obtain required density before the mixture cools to 85°C (185°F)
- Finish rolling – Rolling done solely for the improvement of the surface while the mixture is still warm enough to permit removal of any roller marks

Within the first two operations, a sequence must be followed to ensure a mat of specified density, shape and smoothness. The sequence dictates which parts of the mat are rolled first and which last, and it varies for thin and thick layers.

Thin Layers (Lifts) When placing a thin lift (less than 50 mm [2 in.] compacted thickness) in single-lane width or full width, the mixture should be rolled in the following sequence:

1) Transverse joint
2) Outside edge
3) Breakdown rolling, beginning on the low side and progressing toward the high side
4) Intermediate rolling; same procedure as Step 3
5) Finish rolling

When paving a thin lift in echelon, or when abutting a previously placed lane or other lateral restraint, the mixture should be rolled in the following sequence:

1) Transverse joint
2) Longitudinal joint
3) Outside edge
4) Breakdown rolling, beginning on the low side and progressing toward the high side
5) Intermediate rolling; same procedure as Step 4
6) Finish rolling

Thick Layers (Lifts) When placing a thick lift (50 mm [2 in.] or more compacted thickness) in single-lane width or full width, the mixture should be rolled in the following sequence:

1) Transverse joint.
2) Breakdown rolling, beginning 300 to 380 mm (12 to 15 in.) from the lower unsupported edge and progressing toward the high side.
3) Breakdown rolling of outside edge. When within 300 mm (12 in.) of the unsupported edge, the roller should advance toward the edge in approximately 100 mm (4 in.) increments in successive passes.
4) Intermediate rolling, beginning on the low side and progressing toward the high side.
5) Finish rolling.

When paving a thick lift in echelon, or when abutting a previously placed lane or other lateral restraint, the mixture should be rolled in the following sequence:

1) Transverse joint.
2) Longitudinal joint.
3) Breakdown rolling, beginning at the longitudinal joint and progressing toward the outside edge. When within 300 mm (12 in.) of the unsupported edge, the roller should advance toward the edge in approximately 100 mm (4 in.) increments in successive passes.
4) Intermediate rolling, beginning on the low side and progressing toward the high side.
5) Finish rolling.

▶▶ Specific Rolling Procedures

Rolling Transverse Joints When the transverse joint is next to an adjoining lane, the first pass is made with a static steel-wheeled roller moving along the longitudinal joint for a few feet. The surface is then checked with a straightedge and corrections are made if necessary. The joint is then rolled transversely, with 150 mm (6 in.) of the drum width on the newly laid material. This operation should be repeated with successive passes, each covering an additional 150 to 200 mm (6 to 8 in.) of the new mat, until the entire width of a drive roll is on the new mixture.

During transverse rolling, wooden boards of the proper thickness should be placed at the edge of the pavement to give the roller a surface to drive on once it passes the edge of the mat. If boards are not used, transverse rolling must stop 150 to 200 mm (6 to 8 in.) short of the outside edge to prevent damaging it, and the edge must be compacted later during longitudinal rolling.

Rolling Longitudinal Joints When using static steel-wheeled or pneumatic-tired rollers to roll longitudinal joints, only 100 to 150 mm (4 to 6 in.) of the roller width should ride on the newly placed lane on the first pass. The bulk of the roller width should ride on the previously compacted side of the joint. In each subsequent pass, more and more of the roller width is allowed on to the fresh mat, until the entire roller is on the new mixture.

A different procedure is employed with vibratory rollers. The roller drums are extended only 100 to 150 mm (4 to 6 in.) onto the previously compacted lane, with the rest of the drum width riding on the newly placed mixture. The roller continues to move along this line until a thoroughly compacted, neat joint is obtained.

For compaction purposes, longitudinal joints can be categorized into two categories: *hot* and *cold*. Each requires a different compaction procedure.

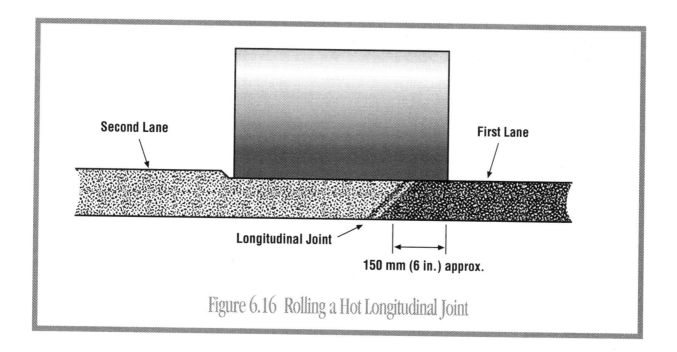

Figure 6.16 Rolling a Hot Longitudinal Joint

Hot Joints – A hot joint is one placed between two lanes at approximately the same time by pavers working in echelon. This produces the best longitudinal joint because both lanes are at or near the same temperature when rolled. The material becomes a single mass under the roller, and there is little or no difference in density between the two lanes. When paving in echelon, the breakdown roller following the lead paver leaves 75 to 150 mm (3 to 6 in.) of the common edge or joint unrolled between the pavers. This common joint is then compacted by the roller following the second paver on his first pass (Figure 6.16). In order to accomplish this effectively, the second paver and roller must keep as close as possible to the first paver to ensure that a uniform density is obtained across the joint.

Cold Joints – A cold joint is one between two lanes, one of which has cooled overnight or longer before placing the adjoining lane. Because of the difference in temperature between the two lanes, there is almost always a difference in density between the two sides of the joint, regardless of the rolling technique used.

In most cases there is a low-density zone at the joint in the lane placed first, and a higher density zone at the joint in the abutting lane. The only practical solution to eliminate this problem appears to be echelon paving or full-width paving. Echelon paving allows the joint to be compacted while the asphalt mixture is hot on both sides. However, since most asphalt paving is done in single lanes, the next best solution is to roll the joint as soon as possible. In any case, longitudinal joints should be rolled as close behind the paver as possible.

Rolling Edges Except in echelon and thick-lift paving, the edges of the pavement should be rolled concurrently with the longitudinal joint. In rolling edges, roller wheels should extend 50 to 100 mm (2 to 4 in.) beyond the pavement edge, provided that lateral displacement of the mixture is not excessive.

After longitudinal joints and edges have been compacted, breakdown rolling should follow immediately.

Table 6.02 Factors Influencing Compaction

Item	Effect	Corrections*
Aggregate		
Smooth Surfaces	Low interparticle friction	Use light rollers; lower mix temperature
Rough Surfaced	High interparticle friction	Use heavy rollers
Unsound	Breaks under steel-wheeled rollers	Use sound aggregate; use pneumatic rollers
Absorptive	Dries mix – difficult to compact	Increase asphalt in mix
Asphalt		
Viscosity		
– High	Particle movement restricted	Use heavy rollers; increase temperature
– Low	Particles move easily during compaction	Use light rollers; decrease temperature
Quantity		
– High	Unstable & plastic under roller	Decrease asphalt in mix
– Low	Reduced lubrication – difficult compaction	Increase asphalt in mix; use heavy rollers
Mix		
Excess Coarse Aggregate	Harsh mix – difficult to compact	Reduce coarse aggregate; use heavy rollers
Oversanded	Too workable – difficult to compact	Reduce sand in mix; use light rollers
Too Much Filler	Stiffens mix – difficult to compact	Reduce filler in mix; use heavy rollers
Too Little Filler	Low cohesion – mix may come apart	Increase filler in mix
Mix Temperature		
High	Difficult to compact – mix lacks cohesion	Decrease mixing temperature
Low	Difficult to compact – mix too stiff	Increase mixing temperature
Course Thickness		
Thick Lifts	Hold heat – more time to compact	Roll normally
Thin Lifts	Lose heat – less time to compact	Roll before mix cools; increase mix temperature
Weather Conditions		
Low Air Temperature	Cools mix rapidly	Roll before mix cools
Low Surface Temperature	Cools mix rapidly	Increase mix temperature
Wind	Cools mix – crusts surface	Increase lift thickness

* Corrections may be made on a trial basis at the plant or job site. Additional remedies may be derived from changes in mix design.

Breakdown Rolling It is important to start the breakdown rolling operation on the low side of the mat (usually the outside of the lane being paved) and progress toward the high side. The reason is that hot mixtures tend to migrate toward the low side of the mat during compaction. If rolling is started on the high side, migration is much more pronounced than if rolling starts from the low side. When adjoining lanes are placed, the same rolling procedure should be followed but only after compaction of the longitudinal joint.

Intermediate Rolling Intermediate rolling should follow breakdown rolling as closely as possible, while the asphalt mixture is still well above the minimum temperature at which densification can be achieved, 85°C (185°F). Intermediate rolling should be continuous until all of the mixture placed has been thoroughly compacted. Regardless of the type of rollers used, the rolling pattern should be developed in the same manner as breakdown rolling.

Finish Rolling Finish rolling is done solely for the improvement of the surface, that is, to remove roller marks so that the surface looks good and rides smoothly. It should be accomplished while the material is still warm enough for removal of roller marks. Vibratory rollers must be operated in the static mode when they are used for finish rolling on pavements that are below 85°C (185°F).

Summary

Compaction is the process of compressing a given volume of HMA into a smaller volume in order to increase the strength and durability of the mixture. It is essential in reducing the permeability of the pavement. Excessive penetration of air and water can be extremely damaging to in-place pavements. Table 6.01 contains a summary of items influencing compaction.

Several factors determine how easily and effectively a mixture can be compacted. Some of these factors are: mixture properties, environmental factors, layer thickness and underlying courses.

Three main types of rollers are commonly used: static steel-wheeled rollers, pneumatic-tired rollers, and vibratory rollers. The number and location of these rollers in the compaction train will vary from job to job. Careful consideration of specific conditions is necessary to determine the proper rolling procedures on each project. Awareness of changes in mixture properties, mixture production rates and environmental conditions is essential in assessing the effectiveness of the compaction process.

The ultimate goal in constructing a quality asphalt pavement is to produce a smooth, strong, uniform, highly durable driving surface. The compaction process is the final opportunity to achieve these results. Knowledge of the variables affecting compaction, along with careful observation, testing and experience, are the most valuable tools in achieving this goal.

| General |

Hot mix asphalt (HMA) mixtures have traditionally been designed using an empirical approach to determine acceptable mix design criteria. Using this approach, mixtures were designed in the laboratory using a specific set of design criteria and were then produced in the field. The pavements constructed using these mixtures were evaluated over a period of time. The actual field performance was used as a guide to judge the adequacy of the original design criteria. If the pavements did not perform to the satisfaction of the agency, the mix design specifications were often adjusted in an attempt to overcome inadequacies in certain performance categories. Although these practices are clearly related, they are often considered separate activities.

The goal of mix design is to arrive at a starting point for establishing process mixing control and uniformity. This starting point is most often referred to as the job mix formula, or JMF. The JMF is the combination of aggregate and asphalt binder materials proposed for use on a project that when tested, using the required mix design procedures, yields results that meet all the established design criteria. For specific mix design procedures refer to the Asphalt Institute's *Mix Design Methods for Asphalt Concrete and Other Hot Mix Types* (MS-2), and *Superpave Mix Design* (SP-2).

It has been documented in many cases that even though a mix is designed for a certain percent of air voids in the laboratory, a plant-mixed sample containing the proper asphalt content, aggregate gradation and compacted by the same laboratory method may show different volumetric properties. The results of FHWA Demonstration Project 74 (Demo 74) indicated substantial differences may exist between the laboratory mix design and the volumetric properties of the field-produced mixture[1]. Most often these changes result in lower air void and VMA values than the target established for the JMF. Figures 7.01 and 7.02 show the differences in air voids and VMA found during the FHWA study.

Significant equipment and material differences exist between the small scale operation of the laboratory mixing bowl and an asphalt mixing facility, which could lead to the mixture property changes noted. Many causes for volumetric changes in field produced mix have been proposed:

- Moisture removed more effectively in the laboratory than in the field
- An excessive amount of baghouse fines returned to the mixing process.
- Highly absorptive aggregates used (field absorption may differ from laboratory absorption)
- Poor sampling used to obtain samples for the laboratory design
- Non-standard mix design procedures used to develop the JMF
- Material sources (location or strata) changed between the mix design and field production
- Aggregate degradation and shape changes during plant mixing.
- Non-uniform stockpiling of aggregates

The changes which occur in the field-produced mixture are likely due to a combination of some or all of the listed causes in any particular case. Attention to detail in material production and mix design development can reduce the potential effects of many of the factors shown.

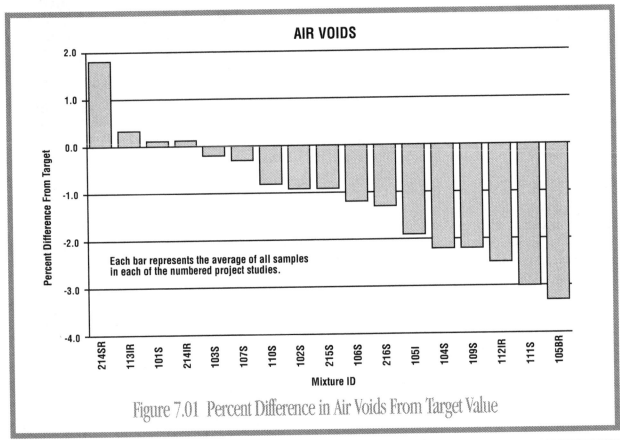

Figure 7.01 Percent Difference in Air Voids From Target Value

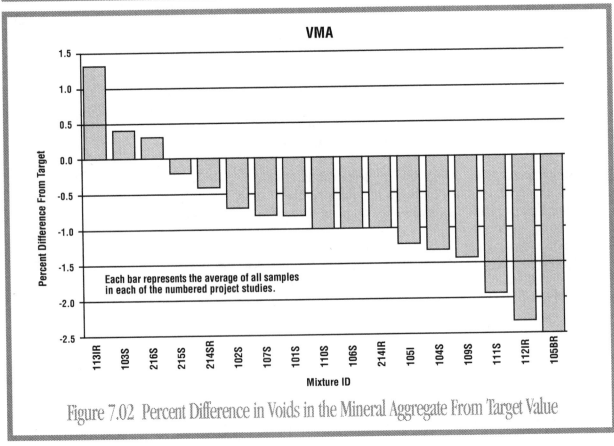

Figure 7.02 Percent Difference in Voids in the Mineral Aggregate From Target Value

QUALITY CONTROL AND ACCEPTANCE OF HOT MIX ASPHALT

Table 7.01 Mix Design Verification Results by Mix Number and Designer[1]		
Study Number	**Mix Designer**	**Mix Verification Result**
101S	Contractor	Go With Changes
102S	Contractor	Go With Changes
103S	Contractor	Go With Changes
104S	State	Redesign
105BR	State	Redesign
105I	State	Redesign
106S	State	Go With Changes
107I	Contractor	Go With Changes
109I	State	Redesign
110S	State	Go With Changes
111S	State	Redesign
1121R	State	Go With Changes
1131R	State	Go With Changes
214SR	Contractor	Go As Is
214R	Contractor	Go As Is
215S	State	Go With Changes
216S	State	Go With Changes

The Demo 74 study also provided information related to the responsible laboratory for developing the initial JMF for the mixtures tested. Table 7.01 shows which mixtures were designed by the state highway agency and those designed by the contractor's laboratory. Changes between laboratory mix design and field-produced mixtures were needed when designed by either responsible party. However, there were no instances where redesign of the mixture was recommended when the contractor had developed the JMF. This suggests that the contractors involved in this study had a good understanding of the materials they used and how to control the volumetric properties of the mixture during production.

With these factors in mind, adjustments can be made by the contractor in the mix design process to accommodate these anticipated changes. For example, if a contractor knows through experience that using a combination of aggregates from certain sources typically results in the return of 1% baghouse fines to the mix during production, this amount of baghouse fines could be incorporated into the aggregate blend during the mix design. By doing this, the volumetric changes from the laboratory design to the field-produced mix could be reduced. A number of other factors contributing to volumetric changes may be known to the contractor. These could also be accounted for during the mix design. It is not reasonable to assume the state agency laboratory personnel would be aware of all the possible factors to consider for all the possible combinations of materials and asphalt mixing plants. Therefore, it is recommended that contractors be responsible for providing acceptable mix designs to the contracting agency prior to mix production.

Field verification of the HMA design is the initial phase of the overall Quality Control (QC) process. It involves testing and analyzing the field-produced mixture to ensure that the criteria established by the specifying agency for the particular mixture are being met. Verification is necessary at the beginning of production of each mix or JMF to measure what differences, if any, exist and what corrective measures need to be taken.

Table 7.02 Sample Quality Control and Acceptance Procedures

I.   Pre-production Sampling and Testing
   A. Aggregate for mix design
   B. Mineral filler/additives, if necessary
   C. Asphalt material from proposed source

II.   Job Mix Formula Approval and Verification
   A. Aggregate gradation
   B. Aggregate physical properties where required
   C. Asphalt Content
   D. Air voids, VMA and VFA
   E. Stability/Strength testing, where applicable
   F. Moisture susceptibility testing

III.   Quality Control Testing During Production by Contractor
   A. Maximum theoretical specific gravity (Rice)
   B. Bulk specific gravity for air voids, VMA and VFA
   C. Aggregate gradation
   D. Asphalt content

IV.   Production or In-place Acceptance Testing by Agency
   A. Asphalt content
   B. Aggregate gradation
   C. Air voids, VMA and VFA
   D. In-place density
   E. Thickness
   F. Smoothness/Ride Quality
   G. Roadway profile

Quality control is one part of a total quality assurance system designed to assure that the quality of the construction and materials conform with the plans and specifications under which it was produced. Activities that occur under the umbrella of this total system are:

- Quality control practices by the contractor designed to monitor the product manufacturing process.
- Acceptance sampling, testing and inspection by the agency to determine if satisfactory quality control has been exercised to attain proper specification compliance
- Independent assurance sampling and testing. This third party involvement is used to provide an independent critique of the entire QA process. Sample quality control and acceptance procedures are shown in Table 7.02.

Field verification is intended to verify that plant production will essentially match the JMF. It often includes increased testing above the minimum specified frequency and can result in adjustments to the JMF or in a complete redesign of the mixture.

A properly designed and administered quality control program for HMA is meant to prevent the production of substandard or out-of-specification material rather than to document the degree of noncompliance after the fact. An adequate quality control process will address the concerns and risks of both the producer and the purchasing agency. The process will provide the producer with confidence and evidence of the product quality, the ability to discern trends in production and anticipate potential problems, and the ability to assess the risk of producing noncompliant materials. The agency will be provided assurance that the materials being produced are within the established criteria, which should result in the desired pavement performance.

Quality Control Testing

Specific properties of the field-produced mixture are measured and compared to the job mix formula and other specification requirements. The quality control (QC) tests that are used will vary depending on the design procedure specified by the controlling agency. Until recently, most organizations that incorporate quality control have used the Marshall mix design method primarily because the compaction equipment is portable, making it economical for quality control techniques. Texas gyratory compaction is also well suited. Since the end of the Strategic Highway Research Program (SHRP), the Superpave system has been used with increasing frequency for mix design and field quality control activities. The Superpave Gyratory Compactor (SGC) was designed for use as a field control compactor as well as for mix design. For information related to the Superpave system refer to the Asphalt Institute's *Superpave Mix Design* (SP-2).

While some tests used in the quality control process are governed by strict test methods, others have more than one alternative option. For a quality control program to operate properly, it is essential that all those involved in the testing activities use standard testing equipment and established testing procedures. When standardization of the equipment and procedures is not employed throughout the quality control process, an increase in test variability will likely occur. This increases the probability that poor materials could be accepted or that quality materials could be rejected.

Figure 7.03 shows the amount of sample variance attributable to each of three factors for the given test. One of the primary goals of a quality control program should be to reduce as much

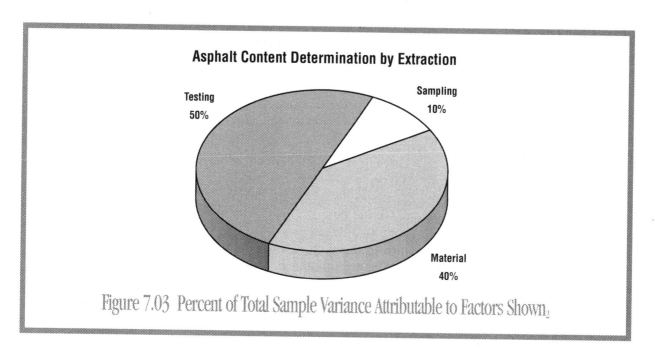

Asphalt Content Determination by Extraction

Testing 50%

Sampling 10%

Material 40%

Figure 7.03 Percent of Total Sample Variance Attributable to Factors Shown

as possible the sampling and testing contributions to the total variance in test results. The allowable variance is generally taken into consideration by providing acceptable plus and minus tolerances to the JMF target values.

Many agencies and industry partners have incorporated technician training and certification programs into the quality assurance process. Both agency and contractor personnel should go through the same training program to ensure all those involved with field quality control and acceptance understand their responsibilities and are qualified to perform their assigned duties. These training activities also address the equipment and procedural requirements of the program.

A standard specification has been established by ASTM to address the testing laboratory and technician capabilities in performing tests on bituminous materials. ASTM D3666, "Minimum Requirements for Agencies Testing and Inspecting Bituminous Paving Materials", discusses the criteria which can be used to evaluate the qualifications of an agency, consultant or contractor laboratory for performing tests on HMA materials. This standard can be consulted when establishing a quality control program.

The following sections discuss the tests used to verify, and subsequently control the mixture's compliance with the job mix formula.

▶▶ Asphalt Content

Many methods can be used to determine the asphalt content of HMA mixtures. The most frequently used method to date is the extraction test, which separates the asphalt and aggregate using a solvent (AASHTO T164; ASTM D2172). This process results in the ability to perform a gradation analysis on the aggregates after the asphalt binder has been removed. "Automatic recordation" can be used during mix production to calculate asphalt content, if the asphalt mixing facility makes detailed measurements of the materials used. Properly calibrated nuclear asphalt content gauges can provide measurements of asphalt content on the produced mixture (AASHTO T287; ASTM D4125), but since the asphalt is not removed, gradation analysis cannot be performed. These gauges grew in popularity due to environmental constraints being placed on the chlorinated solvents commonly used in extraction testing. Asphalt content determination by the ignition method has been increasing in use in recent years. The ignition method uses very high temperatures ±538°C (1000°F) to "burn off" the asphalt binder from the mixture sample. This method also requires that a calibration be performed on the aggregates used in the mixture to account for possible aggregate degradation during the testing process. Due to the potential for aggregate gradation changes to occur during this test, caution is recommended when using the remaining aggregates for gradation compliance testing.

▶▶ Aggregate Gradation

Various ways also exist to determine aggregate gradation. The aggregate cold feed belt or hot bins are sometimes sampled prior to mixing with asphalt. However, testing of the plant-mixed material after extraction or ignition is the only true measurement of the aggregate gradation in the final mixture. Depending on specific aggregate properties, it is not uncommon during production for an additional one half to one and one half percent of minus 0.075 mm (No. 200) material to be returned to the mix from a baghouse emission control system. A wet scrubber system may reduce the amount of fines in the mix by an equal amount. In addition, degradation of the aggregates may occur in the drying and mixing process in a drum mix plant. These changes will not be realized if gradation testing is not performed on aggregate from the plant-mixed material. This amount of change in the fines content of the mixture can have a profound effect on the mixture volumetrics.

➤➤ *Maximum Specific Gravity*

The theoretical maximum specific gravity, G_{mm}, of the bituminous paving mixture (AASHTO T 209; ASTM D 2041) is a key measurement during both laboratory mix design and quality control procedures. Multiplying the G_{mm} by the unit weight of water, (γ_w, will yield the theoretical maximum density of an asphalt mixture. This is the density of the paving mixture in a "zero air voids" condition. Also called the "Rice" specific gravity after its developer, G_{mm} is the ratio of the weight in air of a unit volume of a voidless asphalt binder and aggregate mixture to the weight of an equal volume of water, at a known temperature. Using a partial vacuum procedure to remove entrapped air from a loose mixture, the test determines the volume of the asphalt mix in a voidless state. The weight of the mix sample divided by this volume is the maximum specific gravity of the mixture.

EXAMPLE 1

Theoretical maximum specific gravity; G_{mm} = 2.438
Unit weight of water = (γ_w = 1,000 kg/m³ (62.4 lbs/ft³)

Maximum density = G_{mm} x (γ_w = 2.438 x 1,000 (62.4) = 2,438 kg/m³ (152.1 lbs/ft³)

The theoretical maximum specific gravity is used to calculate the percent air voids of laboratory compacted samples. The maximum theoretical density is used to calculate the relative density of field compacted pavement cores. The relative density referred to is the ratio of the density of the pavement cores to the maximum theoretical density, not to the laboratory density of compacted mixture samples.

When conducting asphalt mix designs, and testing the mixture properties in the field, it is essential to take asphalt absorption into account. The amount of asphalt absorbed into the aggregates during production, paving and compaction of the mixture will be a function of several factors:

- Absorption characteristics of the individual aggregates and the aggregate blend
- Temperature of the mixture
- Amount of time mixture is maintained at elevated temperatures

It is difficult to determine the amount of asphalt absorption that will ultimately occur in the roadway through laboratory testing. Estimations can be obtained through experience with specific aggregates and knowing the water absorption characteristics of the materials. However, it has been demonstrated that the G_{mm} value changes as the asphalt absorbed into the aggregate varies with a given mixture. This can be explained by recalling the weight/volume relationship of material when determining its theoretical maximum specific gravity.

Immediately after mixing a sample of HMA, using known weights of aggregate and asphalt binder, the volume of the uncompacted mixture can be determined using the Rice test mentioned above. We could assume that a certain amount of asphalt absorption had taken place during this procedure. If the mixture is held at a high temperature for a longer period of time after mixing, the asphalt binder will continue to be absorbed into the pores in the aggregate. The assumption could also be made in this instance that the relative absorption would be greater than in the first case. The weight of materials remains constant from one example to the other. However, as the asphalt absorption increases, the overall volume of the mixture sample

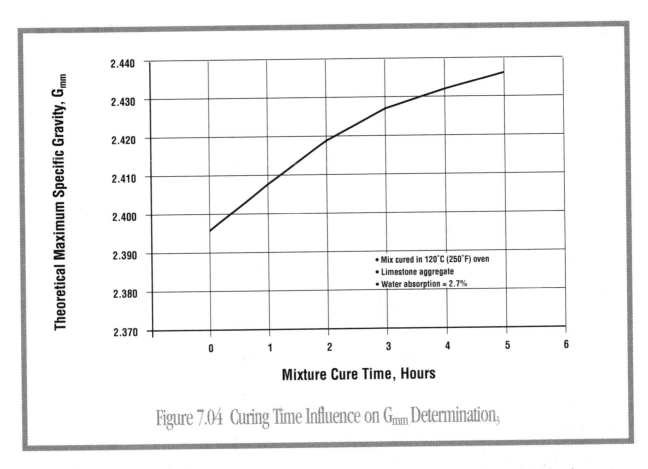

Figure 7.04 Curing Time Influence on G_{mm} Determination,

decreases. This results in a higher G_{mm} value relative to the first test result. For highly absorptive aggregates, this potential variability in the G_{mm} value is greater than with aggregates having lower absorption characteristics. Figure 7.04 shows the effect of cure time on G_{mm} test results for a particular aggregate blend.

It is important to maintain elevated temperatures in quality control testing as close as possible to the amount of time used for development of the mix design. This mix design consideration is often referred to as "curing" the mixture. Many agencies have developed standard curing times between two and four hours for mix design practices. To avoid using improper G_{mm} values for quality control decisions, it is recommended the same curing time be used for field testing.

►►Bulk Specific Gravity

A sample of the plant-produced mixture is cured and compacted using the same procedure used in the mix design (such as a specific number of Marshall hammer blows or number of gyrations). The compacted sample is then used to determine the bulk specific gravity, G_{mb}, of the hot mix asphalt (AASHTO T 166 or T 275; ASTM D 1188 or D 2726). Multiplying the G_{mb} by the unit weight of water will yield the bulk density of the compacted sample.

EXAMPLE 2

Bulk specific gravity = G_{mb} = 2.344
Unit weight of water = (γ_w = 1,000 kg/m^3 (62.4 lbs/ft^3)

Bulk density = G_{mb} x (γ_w = 2.344 x 1,000 (62.4) = 2,344 kg/m^3 (146.3 lbs/ft^3)

➤➤ *Air Voids*

Since G_{mb} is measured on the compacted mixture specimen, the measured volume includes air contained within the sample. The percent air voids, V_a, of the compacted mixture is expressed as a percentage of the total bulk volume of the sample and is calculated using the bulk and maximum theoretical specific gravity in this equation (AASHTO T 269, ASTM D 3203):

$$V_a = [(G_{mm} - G_{mb})/G_{mm}] \times 100$$

EXAMPLE 3

$$G_{mm} = 2.438 \qquad G_{mb} = 2.344$$

$$V_a = \frac{(2.438 - 2.344)}{2.438} \times 100 = 3.9 \text{ percent}$$

EXAMPLE 4

Assume the G_{mm} value used in Example 3 was determined after a curing period of three hours. Also assume that after the same mixture was cured for a period of one hour a G_{mm} value of 2.419 is obtained. The G_{mb} of 2.344 is the same in both examples.

$$V_a = \frac{(2.419 - 2.344)}{2.419} \times 100 = 3.1 \text{ percent}$$

The difference in the air void calculation between Examples 3 and 4 illustrates the importance of allowing the mixture to cure in the field for the same period of time as in the mix design. An air void result of 3.1% might cause the QC supervisor on a project to reduce the asphalt content to raise the air voids to a production target of 4%. However, by not accounting for the continuing asphalt absorption taking place, the result would yield a mixture with an asphalt binder content which is too low for this particular mixture.

➤➤ *Stability and Flow*

Marshall stability and flow properties can be measured on the laboratory compacted samples of field-produced material, and some agencies include them in their quality control testing requirements. However, the reliability of stability and flow as quality control tests is less than density/voids analyses, since Marshall stability and flow values are affected by many different aggregate and asphalt properties. The values obtained are not necessarily an indication of adequate mixture performance. The engineering properties of asphalt mixtures are better defined by the volumetric proportions of asphalt binder, aggregate and air contained within the compacted mixture. It is believed that if the volumetric properties of air voids and VMA, along with the asphalt content and gradation, of the mixture are properly controlled, then stability and flow will correspondingly meet the appropriate specifications.

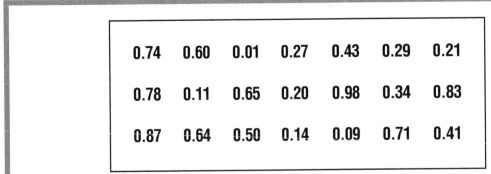

Figure 7.05 Example Excerpt From a Random Number Table

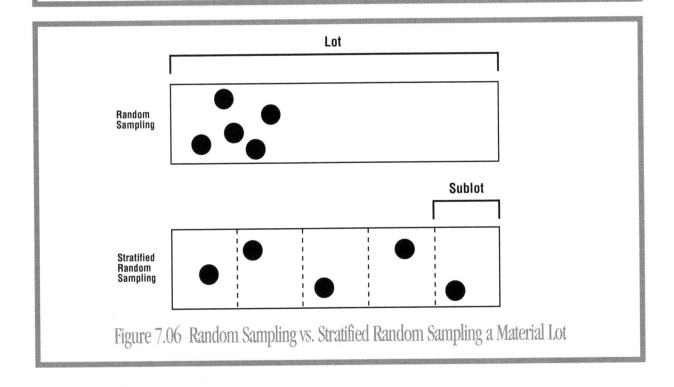

Figure 7.06 Random Sampling vs. Stratified Random Sampling a Material Lot

Sampling and Testing Plan for Quality Control

For a quality control program to be successful, the samples obtained for testing must be representative of all the materials produced for the production run. Samples must be taken often enough to ensure that the material is being produced uniformly throughout the entire production process. For quality control testing of hot mix asphalt, samples are taken at designated intervals. The actual sampling interval, referred to as the sampling frequency, will depend on plant production rates and overall project size.

Typical quality assurance sampling and testing programs use a random sampling procedure to identify where and/or when samples are obtained for testing. A truly random sampling process ensures that any specific increment of a quantity of material has an equal probability of being selected. Random sampling also means that no bias is introduced into the selection process. An effort to include some obviously deficient or passing materials would not constitute a random process.

The most commonly used method of obtaining random samples is by using a random number table (see Appendix C). The table is entered by some predetermined means which is also random in nature. The required group of numbers can be selected by any process. From the starting point, rows or columns could be chosen. Alternate numbers or consecutive numbers can be chosen. Any desired format can be used to select the group of numbers needed. Computers, and some calculators, can also be used to generate random numbers. Figure 7.05 shows an example of an excerpt from a random number table.

The production of HMA is divided into segments of relatively large size which are tested for specification compliance. These large quantities are referred to as a "lot." A lot can be defined as a measured amount of material assumed to be produced by the same process. Sampling of construction materials can be based on divisions of time, area paved, distance paved, weight produced (tonnage), compacted volume or any other suitable increment. For purposes of discussion, assume the lot is based on tonnage of HMA. A lot is often the total tonnage of a single day's mixture production or a specific tonnage of material. Since it is not practical to test every metric ton (ton) of material in the lot, samples are chosen and tested which are assumed to represent the entire lot of material. This would only be true if the process through which the HMA was produced was consistent from the start to the end of the lot.

Since random sampling procedures provide for an equal chance of any increment in a given lot being selected, the possibility exists that all samples taken may be grouped tightly together. For example, if 5000 metric tons (5500 tons) of HMA comprise a lot, it is possible that all samples could be obtained from the first 2000 metric tons (2200 tons) produced. The assumption could be made that the test results obtained represent the total lot tonnage. However, it is desirable to distribute the sampling process throughout the entire lot.

This is accomplished by dividing the lot into equal divisions and randomly selecting locations within these divisions for testing. These smaller divisions are referred to as sublots. In the previous example, the 5000 metric tons (5500 ton) lot could be divided into five equal sublots of 1000 metric tons (1100 tons) each. The random sampling process would then be used to sample each of the successive sublots of material. This process ensures that the sampling and testing is being "spread out" over the entire lot.

The process of dividing the lot into equal divisions and randomizing sample locations inside these sublots is termed a "Stratified Random Sampling" process. The sampling schedule for a particular project would define the quantity of material included in the sublots, as well as the lots. Figure 7.06 illustrates the two possibilities. The top half of the figure shows the possible result if the entire lot were sampled on a random basis. Note the tight grouping of the samples at the beginning of the lot. The second half of the figure shows the stratified sampling procedure resulting in more evenly distributed sample locations.

EXAMPLE 5

Your project specifies that a lot of HMA consists of all the tonnage produced in a single day. The projected total is 5000 metric tons (5500 tons) for the day. Specifications also require the lot to be divided into equal sublots of 1000 metric tons (1100 tons) each. Using the excerpt from a random number table in figure 7.05, determine the sampling locations for the day's production. (Your random number procedure has placed you at the upper left corner of the portion of the table shown, and it specifies that you move horizontally to the right.)

Total number of sublots required = 5000/1000 = 5 sublots

Five random numbers are needed. The numbers chosen are: 0.74, 0.60, 0.01, 0.27 and 0.43

Sublot		Sample Metric Ton (Ton)
#1: (0-1000 metric tons)	1000 x 0.74 = 740	740 (814)
#2: (1001-2000 metric tons)	1000 x 0.60 = 600 + 1000 = 1600	1600 (1760)
#3: (2001-3000 metric tons)	1000 x 0.01 = 10 + 2000 = 2010	2010 (2211)
#4: (3001-4000 metric tons)	1000 x 0.27 = 270 + 3000 = 3270	3270 (3597)
#5: (4001-5000 metric tons)	1000 x 0.43 = 430 + 4000 = 4430	4430 (4873)

It would be impractical to expect that the exact calculated ton of material would be sampled in the above example. It is relatively easy to determine which truck load of mix contains the "sample ton" through review of the weigh tickets provided at the plant. The calculated "sample ton" would be taken from the materials contained in that specific truck either out of the truck box itself, behind the paver, from the paver hopper or another location as indicated in the sampling and testing plan.

EXAMPLE 6

The same project in Example 5 also requires density testing on cores taken from the compacted pavement. Assume that the length of the pavement constructed is to be divided into five equal areas (sublots) for density testing. The total length of roadway paved for the day (one lot) was 10370 meters (34000 feet). The pavement width is 3.6 meters (12 feet), and a single lane was paved. Using the same random number procedure as above for rows two and three in figure 7.05, determine the longitudinal distance from the start of paving and offset from centerline of the cores.

The total length paved = 10370 meters (34000 feet)
Sublot length = 10370/5 = 2074 meters (6800 feet)
Longitudinal distance random numbers (row 2) = 0.78, 0.11, 0.65, 0.20, and 0.98
Centerline offset random numbers (row 3) = 0.87, 0.64, 0.50, 0.14 and 0.09

Core #1: (0 to 2074 m from start of paving)
Distance = 2074 x 0.78 = 1618   1618 m (5307 ft) from start
Centerline offset = 3.6 x 0.87 = 3.1   Offset = 3.1 m (10.2 ft)

Core #2: (2074 to 4148 m from start of paving)
Distance = 2074 x 0.11 = 228   228 + 2074 = 2302   2302 m (7552 ft) from start
Centerline offset = 3.6 x 0.64 = 2.3   Offset = 2.3 m (7.5 ft)

Core #3: (4148 to 6222 m from start of paving)
Distance = 2074 x 0.65 = 1348   1348 + 4148 = 5496   5496 m (18030 ft) from start
Centerline offset = 3.6 x 0.50 = 1.8   Offset = 1.8 m (5.9 ft)

Core #4: (6222 to 8296 m from start of paving)
Distance = 2074 x 0.20 = 415   415 + 6222 = 6637   6637 m (21775 ft) from start
Centerline offset = 3.6 x 0.14 = 0.5   Offset = 0.5 m (1.6 ft)

Core #5: (8296 to 10370 m from start of paving)
Distance = 2074 x 0.98 = 2033   2033 + 8296 = 10329   10329 m (33887 ft) from start
Centerline offset = 3.6 x 0.09 = 0.3   Offset = 0.3 m (1.0 ft)

ASTM D3665, *Standard Practice for Random Sampling of Construction Materials*, provides information and procedures for obtaining unbiased material samples. This standard can easily be adopted for use in sampling HMA mixtures. Random number tables are also provided in this standard.

| **Data Analysis** | Quality control involves two different levels of analysis performed on the HMA. The first involves field verification, the analysis of the mixture |

on the first day or two of full production, to compare the mixture to the job mix formula. The second uses day-to-day quality control tests performed to determine if the mixture properties have exceeded production tolerance limits.

►►Job Mix Formula Verification

At the beginning of production, asphalt content, gradation, and mixture volumetric analysis tests are performed to compare field-produced mixture properties with the job mix formula. These tests will indicate if the aggregate characteristics have varied from those used in the mix design, and may indicate if problems exist from possible changes in the aggregate after processing through the dryer.

At this point, the field verification results may show that changes in the mixing process are necessary to meet the job mix formula. For example, minor changes in the asphalt content may bring a mixture back within the tolerances of the volumetric requirements. Alternatively, if the mixture is meeting overall agency specifications but not the mix design targets, the job mix formula can be adjusted to accept these new targets. Finally, any dramatic differences between the laboratory design and field-produced mixture may necessitate a new mix design using the actual production materials.

►►Quality Control Testing

Once the job mix formula has been verified, daily quality control testing can provide an early warning of potential problems by indicating if the mixture properties are near the specification limits or deviate from the specifications. This daily testing is a part of plant process control that can identify potential problems before many tons of mix have been placed in the field.

Daily quality control test values are plotted on control charts. Continuous plots of mix data such as percent air voids, VMA, asphalt content, and aggregate percentages passing certain sieves such as 4.75 mm, 0.600 mm and 0.075 mm (No. 4, No. 30, and No. 200) provide a graphical representation of the production process. Target values and upper and lower control limits are set for each material property. The target value is the value specified by the job mix formula or the appropriate specification criterion, and the upper and lower control limits are the job mix formula or specification value plus and minus the allowable tolerance, respectively. The production values plotted in relation to these limits can be used to analyze the results of mixture production and make necessary adjustments to keep the production process within specification limits.

Figure 7.07 shows a set of control charts of asphalt content during production. The top chart shows the value of each asphalt content test performed. The bottom chart shows the moving, or running average of the asphalt content data. The moving average is calculated from consecutive tests values, typically three to five values per subgroup. After each test is performed, the new test value replaces the oldest test value in the subgroup to calculate the new moving

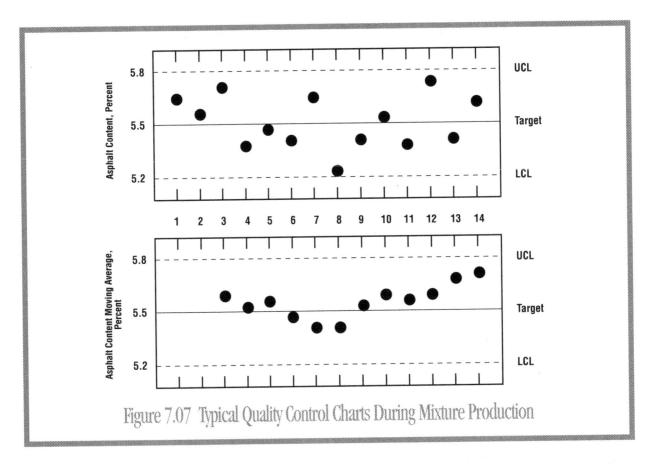

Figure 7.07 Typical Quality Control Charts During Mixture Production

average. A line is typically drawn to connect the points on the chart. The moving average values can be plotted on a separate chart or on the same chart as the individual test values. If the two are plotted on the same chart, a different colored line would normally be used to connect the test values. For example, a black line could be used to connect the values for the individual test results and a red line for the moving average values.

When analyzing quality control data, it is important to recognize sources of variation in the data. These sources include variation in the testing and sampling procedures, normal variations in the materials and production process, and variations due to problems in production. Following the testing and sampling procedures exactly as specified, and as presented in the appropriate training program, will help minimize this variation. Obviously, adjusting the production process on the basis of erroneous test results is not desirable.

The control charts can help differentiate between variation inherent in the material and production variation. They can also provide early signals of potential problems needing attention. The test data should be dispersed randomly about the target value and between the control limits. A few possible indications of existing or upcoming problems are:

- Values consistently higher or lower than the target
- Gradual or erratic shifts in the data
- Systematic cycling of the data

The moving average control chart in Figure 7.07 provides an indication of the overall "trend" of the process. It would be more reasonable to react to the process trend as opposed to making decisions based on individual test results. From point 11 to 14 there is an apparent trend for the average asphalt content of the mixture toward the upper end of the specification range. If this trend were to continue, the HMA supplier runs the risk of producing non-specified materials. This trend would indicate the need for some adjustment in the process to move toward the target asphalt content.

Acceptance Criteria

The particular criteria by which HMA construction projects are accepted depend on the type of specifications used for the project. There are various types of specifications in use, or being proposed, which emphasize different control parameters. They are:

- Method, or recipe specifications
- End result specifications
- Quality assurance specifications
- Performance based specifications

Method specifications spell out specifically the things or procedures that are used in the construction process. Equipment types and quantity, materials criteria such as aggregate gradations, method of construction such as number of roller passes, and other items, very closely control the construction process.

End result specifications move away from controlling the process of construction and concentrate more heavily on the testing of the mixture and pavement after completion for specification conformance. With this type of specification it is critical that the criteria used to judge the test results be adequate for pavement performance.

Quality assurance specifications are based on statistical sampling and testing performed by the contractor and the purchasing agency. Quality control procedures conducted by the contractor are used to assure the specifications are being met throughout the process. Acceptance sampling and testing, performed by the agency, checks the contractor's QC process and determines if the materials should be accepted. Independent assurance sampling and testing is performed by a third party to make sure the entire process is performing adequately. This is typically required only on federal aid projects but may be specified by other agencies.

Performance based specifications emphasize performance of the finished product over time and do not concentrate on construction activities. Specific performance criteria are established in the specification which the finished roadway must meet over a specified length of time. Warranty specifications would fall under this category. In addition, specifications which use performance related tests to evaluate the ability of a mix to perform as intended over a period of time are considered to be performance based specifications.

▶▶ Traditional Acceptance Plans

Acceptance criteria are usually based on a number of specific properties of the mixture and the finished pavement. Traditionally, the following set of items were used to evaluate HMA construction projects.

- *Materials* – Aggregate gradation & quality, asphalt binder properties, asphalt content and mixture volumetric properties
- *Thickness tolerances* – Cores, yield checks, string line or paver skis
- *Pavement smoothness* – Straight edge, profilograph or profilometer
- *Pavement density* – Cores and nuclear gauge readings

For many years, much of the responsibility for the traditional methods of acceptance was held by the specifying agency. Contractors often times submitted proposed aggregate and asphalt materials to the agency, which would then perform a mix design. At the start of production, the agency would conduct asphalt content tests and aggregate gradations to make sure the job mix formula was being duplicated within allowable tolerances. As long as the job mix formula was adhered to, the assumption was made that the mixture volumetric criteria were also being met. As stated earlier, this was often times a mistaken assumption.

This type of combined process control and acceptance also required the agency to be responsible for making adjustments to the aggregate proportions and/or the asphalt content of the mixture. By changing the mixture proportions at the plant, the agency actions would potentially have a direct effect on the contractor's ability to achieve density in the finished pavement. Often times penalties were assessed to the contractor for low densities and the contractor was not in a position to be able to make changes to the mixture to remedy the situation.

For these and other reasons, process control has been incorporated into the quality assurance type of specifications. Here the contractor is more in control of the finished product. Changes made at the asphalt plant which affect the placement and compaction operations are made by the contractor.

►►Volumetric Mixture Control

The recent trend toward a quality assurance type of specification has emphasized mixture volumetric properties rather than the combination of individual material components. Mix design methods used today focus on volumetric properties such as air voids, VMA and voids filled with asphalt. These properties are much better indicators of the engineering properties of the mixture than are asphalt content and aggregate gradation.

These volumetric properties must be controlled in the field as well as be established in the mix design. Nevertheless, gradation and asphalt content determinations are also important to enable educated adjustments to the production process. Judgments on the quality of the finished product which are based on volumetric properties will lead to greater performance of the pavement.

The quality of the finished pavement is not entirely based on mixture properties. The density of the roadway after construction plays a major role in the overall performance of the roadway.

Density Specifications

Quality control of the HMA involves testing and analyzing the field-produced mixture to ensure that the mix design criteria established for the particular mixture are being met. In most cases, pavement density specifications are used to judge the acceptability of the compaction process during construction.

The goal of compacting a hot mix asphalt pavement is to achieve an optimum air void content and provide a smooth, uniform surface. The resultant, in-place air void content of the HMA is probably the single most important factor that affects performance of the mixture throughout the life of the pavement.

The activities involved with the proper design, production, placement, and compaction of the asphalt mixture are all combined to achieve the in-place density of the HMA pavement and ultimately determine whether the pavement will perform as expected. The density specifications to which the pavement is built are used to stipulate the acceptable level of compaction achieved.

A typical density specification represents a comparison between the in-place density of the pavement that is achieved after final compaction, and a reference density. One of three reference densities is typically used in density specifications: Laboratory density; maximum theoretical density; or control strip density. Use of the maximum theoretical density to determine HMA pavement density compliance is preferred by the Asphalt Institute.

➤➤ *Laboratory Density*

This method compares in-place density to a laboratory-compacted sample of field-produced asphalt mix, and is particularly applicable to Marshall compaction procedures. The Superpave method of mix design uses the Superpave Gyratory Compactor (SGC) for laboratory compaction. It has been shown that the SGC procedure is also a method which is well adapted to field laboratory compaction operations. With either method, a reference density is established to which the density of the compacted pavement is compared. The field-produced HMA is compacted using the same compactive effort used during the mix design (e.g. 50 or 75 blows for Marshall compaction or design number of gyrations with the SGC) and the laboratory density is measured using the bulk specific gravity test.

In terms of specification compliance, an agency compares the in-place core density, or nuclear density readings, to the reference density in the form of a ratio:

$$\text{Percent of Laboratory Density} = \frac{\text{In-Place Density} \times 100}{\text{Laboratory Density}}$$

When it has been verified that the field-produced mix matches the mix design volumetric properties, the laboratory compacted samples should provide the same air void content used in the mix design. This is typically four percent. If an in-place air void content of 8 percent is desired for a mix designed at four percent voids, the in-place density should be 96 percent of the reference laboratory density.

➤➤ *Maximum Theoretical Density*

The maximum theoretical density provides the unit weight of the mixture as if it were compacted to a zero air void condition. Using the Rice test method (AASHTO T 209, ASTM D 2041), the maximum theoretical density of the field-produced mixture is determined as the reference density. The relative density of the in-place pavement is again calculated as the ratio of the in-place density to the reference density, which in this case is the maximum theoretical density:

$$\text{Percent of Maximum Theoretical Density} = \frac{\text{In-Place Density} \times 100}{\text{Maximum Theoretical Density}}$$

Since the maximum theoretical density represents a voidless mixture, an in-place air voids content of 8 percent will always be 92 percent of the reference maximum theoretical density, regardless of the mix design air voids value.

To obtain meaningful results, the field produced mixture samples must be cured to the same extent as was done during the mix design process. If the loose mixture samples are not adequately cured in the field, the target maximum theoretical density will be artificially low due to the relatively low asphalt absorption (resulting in greater mixture volume) which has occurred. Under this scenario, the in-place pavement density (and its inverse, pavement air voids) could be determined to be acceptable when in fact, the actual voids were substantially higher. This situation could lead to premature pavement deterioration, and it therefore illustrates the importance of proper curing of the field samples.

▶▶ Control Strip Density

This process calls for the construction of a pavement control strip, also called a test strip, of a minimum length or tonnage of mix at the start of each pavement course being laid. The control strip is part of the paving project. A new control strip should also be constructed if major changes in mixture production or placement occur. A nuclear density gauge is typically used to monitor the densification process. A nuclear reading is taken at one or more locations on the mat after successive passes with each roller. When the maximum density of the control strip is achieved, the compaction process is complete. Maximum density is said to have been achieved when the increase in density after successive roller passes is less than 16 kg/m³ (1 lbs/ft³), or at some other value determined by the agency. After compaction of the control strip is completed, a specified number of bulk specific gravity (density) tests are measured on core samples taken from random locations within the control strip and averaged to obtain the reference density. The cores are also used to calibrate the nuclear gauge if further density control will rely on the nuclear readings. The reference control strip density must then be compared to either the laboratory or maximum theoretical density of the field-produced HMA to determine if densification is adequate and acceptable. Even though its maximum density was achieved during control strip construction, this density may not be at an acceptable level for good pavement performance. Several factors can affect the maximum density achieved during the placement of the control strip:

- Aggregate gradation
- Amount of crushed particles
- Asphalt binder content
- Mix temperature
- Weather conditions
- Number and types of rollers
- Material beneath the control strip

The combination of all the referenced factors must be controlled during the construction of a control strip. Manipulation of one or a combination of factors during production paving should not be allowed. It is also important that the material or pavement course beneath the control strip be essentially the same as the remainder of the area to be paved.

Once an acceptable control strip has been obtained, since the in-place density is exactly the reference density, typically 98 to 100 percent of the reference density is the desired average target density during construction. Also, individual test results should typically be no less than 95 to 96 percent of the target density.

▶▶ Method Density Specification

A "method" specification, sometimes referred to as "ordinary compaction", has no reference density against which the in-place density and air voids are compared. This type of specification contains items such as number and type of rollers to be used, number of passes of each roller, use of temperature measurements, descriptions such as "surface is rolled until free of roller marks," etc. Judgment is the primary decision tool for determining optimum compaction when using this type of specification. Method specifications are generally only applicable for smaller projects with light traffic, areas inaccessible to standard compaction equipment, or thin lift construction (25 mm [1 in.] or less), such as leveling courses and thin HMA overlays. In these cases, cost and the inability to obtain meaningful data from thin, in-place pavements preclude the use of a reference density specification.

➤➤ *Reference Density Specifications*

Satisfactory pavement performance resulting from the use of any reference density specification depends on such factors as:

- Properly designed and plant produced mixtures.
- Proper sampling and handling procedures of the loose samples from the mixing facility or roadway.
- Proper field laboratory testing procedures, especially correct compaction techniques and maximum theoretical density as appropriate.
- Proper sampling, handling, and testing of the pavement core samples.
- Adequate field confinement of the mixture during the compaction process.

The relationship between the reference density measurements and the air voids of the in-place pavement is shown in Figures 7.08, 7.09 and 7.10. An in-place air voids target equal to 8 percent is depicted against each type of reference density. Eight percent is selected here because it is believed that if this level of compaction is achieved at the time of construction, four percent air voids will be achieved in a few years after further densification of the pavement under traffic. Some agencies may prefer a target less than eight percent air voids in the compacted pavement. A target higher than eight percent is not recommended.

It should be noted that while the comparison between maximum theoretical density and in-place air voids content is a consistent one, the relationships between the other two reference density types and in-place air voids will shift up or down depending on the actual mix design and compaction criteria used in the specification. For example, if the mix design air voids is five percent as shown in figure 7.8, then 100 percent of laboratory density would be at five percent air voids. The required percent laboratory density to achieve 8 percent in-place air voids is 97 percent. If a compaction criterion of 96 percent of laboratory density were used, an in-place air void content of nine percent would result (not 8 percent).

Similarly, for a mix design air void content of 3 percent, as shown in figure 7.10, a compaction criterion of 95 percent of laboratory density is required to achieve in-place voids of eight percent. The compaction criterion mentioned in the previous example (96 percent) would result in seven percent in-place voids.

It is important to understand the relationships between the mix design air void content and its effect on reference density specifications. A particular density specification may provide satisfactory field compaction on one project, but due to changes in mixture design criteria, may be inadequate on another.

The use of reference density specifications (laboratory compaction, maximum theoretical and control strip) is appropriate for all projects with a lift thickness greater than 25 mm (1 in.). Each of the reference density specification procedures have additional considerations that may make one more favorable than another on a particular project. These considerations include traffic volume, subgrade support, size of the project, construction and testing schedules, and any lift thickness variation.

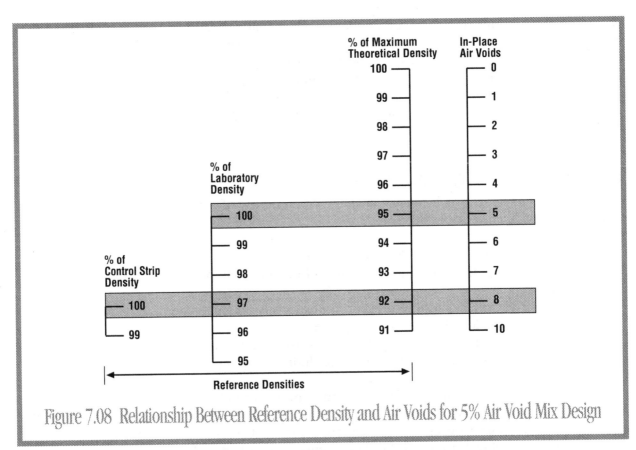

Figure 7.08 Relationship Between Reference Density and Air Voids for 5% Air Void Mix Design

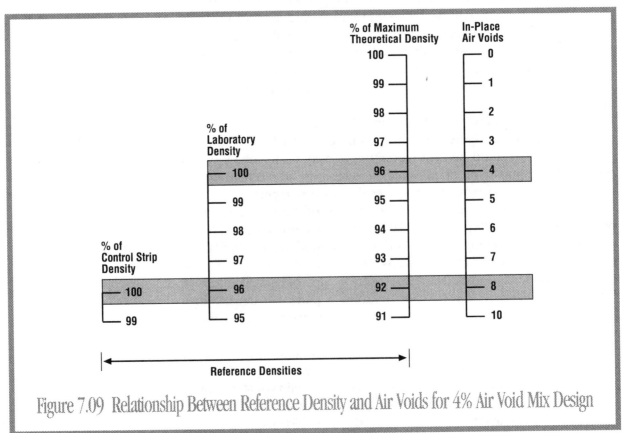

Figure 7.09 Relationship Between Reference Density and Air Voids for 4% Air Void Mix Design

QUALITY CONTROL AND ACCEPTANCE OF HOT MIX ASPHALT

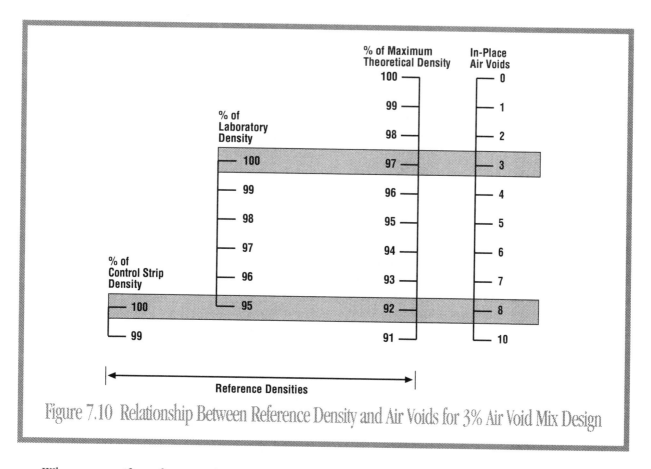

Figure 7.10 Relationship Between Reference Density and Air Voids for 3% Air Void Mix Design

When a specific reference density procedure (laboratory, maximum theoretical, or control strip) is chosen for a project, this same reference density process should be used throughout the testing and determination of in-place density. This will ensure that a valid comparison exists in the determination of density compliance. A higher degree of compaction monitoring is necessary in the initial stages of the construction process, regardless of the density specification used, to ensure optimum results from the compaction process.

In addition to minimum compaction, it is also necessary to avoid too much compaction. If high density (below 3 percent air voids) in a compacted mix is noted, the cause should be determined and corrected. This problem may also require that the mix be adjusted or redesigned. Pavements with an initial air void content below three percent are susceptible to permanent deformation and flushing after further consolidation due to traffic. Conversely, high initial air void levels, above eight percent, are likely to yield pavements which age prematurely. These low density pavements are more permeable to the damaging effects of air and water migrating into the interior of the mat. The result is a brittle pavement which is more likely to crack due to age hardening of the asphalt binder and effects of thermal expansion and contraction. These pavements are also more susceptible to stripping of the asphalt film from the surface of the aggregate, resulting in a reduction in pavement strength.

For these reasons it is recommended that compaction specifications require the resulting in-place air voids, immediately after construction, be between four and eight percent (96 to 92 percent of maximum theoretical density). Studies have shown this range in pavement air voids offers adequate resistance to air and water intrusion while allowing consolidation under traffic to take place without severe permanent deformation or flushing.

Summary   Based on the present knowledge of plant production and pavement behavior, quality control must be utilized to manage the process of asphalt mixture production in order to minimize the variability between mix design goals set in the laboratory and actual mix results achieved in the plant.

A statistically based sampling and testing program for quality control does not eliminate the need for adequate agency field inspection. Obviously defective materials should be considered for rejection, even though they may not necessarily be chosen for testing under the quality assurance process. For example, suppose a sublot is defined as containing 1000 metric tons (1100 tons) of HMA and the trucks delivering the mix to the project are averaging 20 metric tons (22 tons) per load. If the specifications require one sample of HMA be obtained for quality control testing per sublot, there is only a 2% chance that any single truck load would be chosen for testing. A properly trained field inspector would need to have the authority to recommend rejection of defective loads of HMA regardless of whether or not they are included in the test results. A defective load would have one or more obvious visual defects, such as contamination, excess asphalt, severe segregation or a temperature deficiency.

The in-place air voids of the HMA after compaction is probably the single most important factor in the acceptance procedure that affects performance of the mixture throughout the life of the pavement. However, specifying compaction is not sufficient for ensuring the success of a paving project or ensuring a durable, long lasting pavement. Compaction specifications are the final step in the total quality assurance of the HMA construction process. Proper mix design, production, quality control, construction and acceptance procedures must be integrated within the project requirements to achieve a quality product.

(1)   D'Angelo, J. and T. Ferragut, "Summary of Simulation Studies From Demonstration Project No. 74 Field Management of Asphalt Mixes," Proceedings, Association of Asphalt Paving Technologists, Vol. 60 (1991) pp. 287-306.
(2)   Bureau of Public Roads. V35, Report #9.
(3)   Martinez, F. and Bayomy, F., "Selection of Maximum Theoretical Specific Gravity for Asphalt Mixture Design," Transportation Research Record 1300, 1991, p.17.

Introduction

It is desirable to conduct all hot mix asphalt (HMA) paving operations without segregation of the mixture. Due to many factors, it is very difficult to eliminate segregation completely. The next best alternative is to build HMA pavements with the least segregation possible. In order to do that, it is necessary to understand the nature of segregation in HMA, as well as its causes and the preventive measures to be taken. Methods of recognizing and avoiding segregation in production and construction of HMA are presented in this chapter. In addition, the analysis of segregated mixes is discussed, and corrective measures are given.

Overview of Segregation

Segregation in HMA is a non-uniform distribution of the various aggregate sizes throughout the mass. The finished mat has a varied texture and more than likely does not meet specification requirements for surface texture, smoothness, or density (Figure 8.01). Segregated areas differ from the approved mix design, and the areas will not meet the volumetric properties required and discussed in *Determining Segregation Level* at the end of this chapter.

Segregation is not a phenomenon new to the HMA industry. It started to occur more commonly in the late 1970's when HMA mixtures experienced an

Figure 8.01 Segregated Hot Mix Asphalt Pavement

increase in the percent of material passing the 0.075 mm (No. 200) sieve and a corresponding decrease in asphalt content. The use of surge and storage bins increased, changes in plants occurred, and trucking increased. Segregation is receiving additional attention at this time for good reason: Agencies are demanding additional life from their pavements. Severely segregated pavements require maintenance sooner than properly constructed ones because of excessive moisture damage, raveling, and premature cracking.

►►Types of Segregation

The most common types of segregation are:

- Random spots which occur intermittently throughout the roadway
- Chevron shaped spots at the beginning and end of truckloads
- Continuous longitudinal streaks at either or both sides of the lane
- Center of paver streaks

Each type of segregation is the result of a specific action in mixing and placing, and the segregation is prone to occur most often at five stages, as shown below. Careful observation and control of mixing and placing operations during theses stages can reduce or eliminate segregation in most asphalt mixes. The economic benefit of reducing or eliminating segregation has been recognized universally, and agencies that specify asphalt pavements are increasingly requiring more to control segregation.

►►Segregation Trouble Spots

Segregation may be caused by the methods used in aggregate handling, and in mixing, storing, transporting and handling the mix, where a condition is created that favors non-random distribution of the aggregate sizes. Years of research and observation have shown that segregation most often occurs at the following stages:

- Mix design (as a factor in segregation potential)
- Aggregate handling and stockpiling
- HMA plant production, such as at the:
    A. Cold bins; loading and feeding
    B. Hot bins
    C. Drum mixer operations
    D. Hot elevator
    E. Pugmill bin gates
    F. Surge and storage bins
    G. Discharge systems
- Trucking operations
- Paver operation

There may be other stages, but these five are the most common and can readily be detected and corrected. Worn or improperly maintained mixing and paving equipment also leads to problems in handling and placing the mix. A technician can visually identify segregation and quantify it from an extraction analysis. The required asphalt content of a mix is directly proportional to the surface area of the aggregate – the finer the mix, the larger the surface area, and the greater amount of asphalt required to coat the aggregate. This relationship can determine whether the mix segregated before or after mixing. The ability to pinpoint the source of the segregation will minimize hours of review work to solve the problem. Detailed evaluations done in the laboratory are described in the section, *Pinpointing Segregation.*

➤➤*Segregation Potential in Mix Design*

Certain characteristics in the makeup of a mix design can increase the potential for segregation in the produced HMA. These characteristics of the mix are the result of mix designers reacting to the realities of modern highway transportation. Increased truck traffic, heavier loads and increased tire pressures have brought about mix designs with a tougher aggregate skeleton. These tougher HMA mixtures may be segregation prone. In addition, environmental restrictions on the amount of fines (dust) that may be released into the atmosphere have resulted in mixes with a higher percentage of material passing the 0.075 mm (No. 200) sieve. This decreases the voids in the mineral aggregate (VMA), therefore reducing the asphalt demand. This reduction in asphalt reduces the cohesion of the mixture, as well as its workability and durability.

The potential for segregation in HMA is increased by:

• Lower asphalt contents
• Higher percentages passing the 0.075 mm (No. 200) sieve
• The use of larger maximum size aggregates
• Coarser and gap-graded gradations within established specification sizes

A lower asphalt content makes the mix prone to segregation, regardless of its gradation. Lower asphalt contents are often the result of trying to achieve volumetric properties, such as air voids, by varying the asphalt content. A low VMA in the aggregate structure and a high percentage of material passing the 0.075mm (No. 200) sieve will result in a low asphalt content, and consequently low cohesion of the mixture.

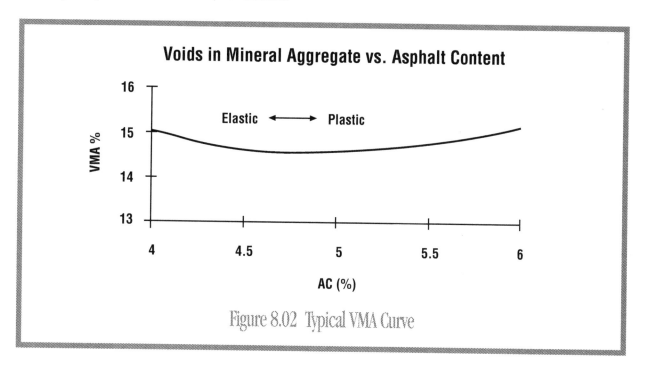

Figure 8.02 Typical VMA Curve

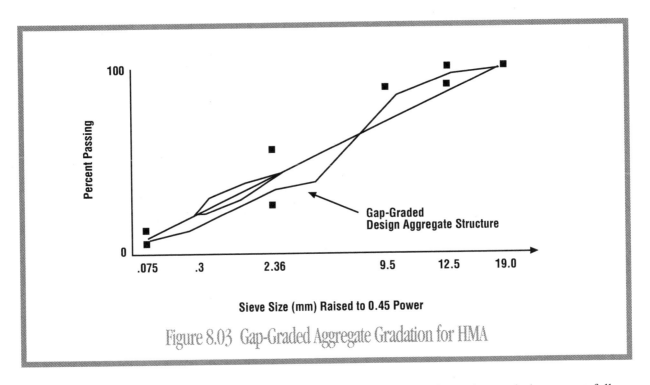

Figure 8.03 Gap-Graded Aggregate Gradation for HMA

A good method to analyze a mix for cohesion is to observe where the asphalt content falls on the VMA curve. For maximum cohesion, while maintaining an elastic mixture, the asphalt content should be slightly to the left of the minimum VMA value.

Figure 8.02 illustrates a typical VMA curve. The minimum VMA value occurs at 4.7 percent asphalt content (AC). The design AC should not be more than one-half of one percent (0.5%) less than the asphalt content corresponding to the minimum VMA value. This is only a rule of thumb, and it should be applied with common sense after evaluating each mixture. Experience has shown that mixtures with asphalt contents below this value are low in cohesion and are prone to segregate. Conversely, asphalt mixtures with a design AC above the value corresponding to the minimum VMA will be over-asphalted and can result in a plastic mixture prone to rutting and shoving.

An optimum asphalt content using these guidelines and Figure 8.02 would be between 4.2 and 4.7 percent. However, the lower limit does not provide enough asphalt to give the mix adequate cohesion and durability. If the design air voids (such as 4.0%) indicates an asphalt content at or near 4.2 percent, the mixture should be redesigned using material with less fines.

The curve in Figure 8.02 shows a typical relationship between VMA and asphalt content of a mixture with a high percentage of material passing the 0.075 mm (No. 200) sieve. The low point of the curve is the critical point of the mixture. This point indicates the value of AC that changes the state of the mixture from elastic to plastic. For maximum cohesion of the mixture – without making the mixture plastic – the asphalt content should be near the maximum of the 4.2 to 4.7 percent range.

Base mixtures with a maximum aggregate size of 50 to 75 mm (2 to 3 inches) are seeing greater use. In addition, some agencies have increased the maximum size of surface and intermediate courses in order to increase resistance to rutting. As the difference between the maximum and minimum size in a mix increases, the tendency to segregate increases. Large stone mixes by their nature are segregation prone.

Another potential cause of segregation in HMA due to its mix design is the gradation. When an aggregate gradation is near the coarse limit of its specification, or when it is gap-graded, the

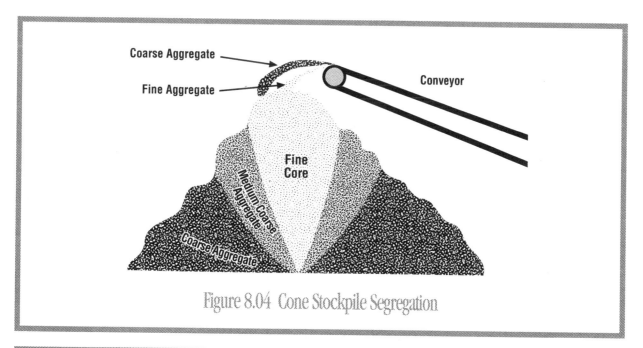

Figure 8.04 Cone Stockpile Segregation

Figure 8.05 Cone Stockpile Built With Aggregate Ladder

potential for segregation is increased. Coarser aggregate gradations are often sought and these sometimes result in gap-graded mixtures. Figure 8.03 shows a gap-graded aggregate gradation.

The use of aggregate blends with diverse specific gravities of the coarser and finer materials also introduces the potential for segregation of HMA. Mixtures with this property tend to segregate during mixing and handling due to the differences in weight of the particles. This is particularly true if the coarse aggregate has a higher specific gravity.

➤➤ Aggregate Handling and Stockpiling

Stockpiling aggregate prior to use allows drainage of water from the aggregate to take place. With such drainage, the stockpile aggregate moisture content approaches a uniform distribution, and control of drying time in the HMA plant is facilitated. Many aggregate plants stockpile by dumping from a truck and pushing aggregate piles with track-type dozers, rubber-tire dozers, or front-end loaders. Some are stockpiled with stacking conveyors that rotate. A single cone stockpile could result in the highest variation of gradation due to segregation. Figure 8.04 illustrates the segregation that occurs in this type of stockpile.

Segregation occurs with movement and through vibration. This is the case when transporting material on the primary conveyor belt, as the finer material moves downward to the belt surface. Figure 8.04 shows that primary conveyor trajectory generates a segregating effect. As the coarser particles are thrust further away from the head pulley, the finer particles drop out closer to it. As the height of the cone of the surge pile increases, the coarser particles tend to roll and slide to the perimeter of the pile.

Segregation becomes more pronounced as the height of a stockpile approaches its peak. Separation of aggregate particles can occur when the aggregate falls a great distance from the conveyor to the stockpile, and degradation can take place as the falling aggregate strikes the stockpile. Both can be minimized by keeping the stockpile at a high level or by the use of boom stackers that can be adjusted vertically to control the height of fall. Another method of building extremely tall stockpiles with stationary conveyors is by the use of an aggregate ladder as shown in Figure 8.05. The ladder handles segregation by keeping the product homogeneous from the head pulley to the point of stockpiling. The drawback to using the ladder may be in an increased tendency to degrade the material.

Stockpiles can be built properly, with minimum segregation, by end-dumping the aggregate from trucks to form piles that are only one layer deep. An example of this operation is shown in Figure 8.06.

Stockpiles of single-size fractions for intermediate and surface mixes are less sensitive to segregation and degradation. Segregation is reduced because the maximum size is less and the ratio of larger to smaller sizes is decreased. Degradation has been reduced because of the inherent beneficiation that has occurred during the processing to a smaller aggregate size. Beneficiation may be defined as the removal or partial destruction of deleterious materials within the aggregate particle.

A fractionating plant is designed to provide separate aggregate piles with a given pile having nearly equal aggregate sizes. A fractionating plant is one of the better processing alternatives for controlling segregation. Blending the various size aggregates together at the loadout point in the hot mix asphalt plant minimizes the effect of segregation and helps to achieve the desired product specifications. Today, most contractors use four or more cold feed bins to provide greater quality control at their HMA plants. Single size aggregates provide greater quality control but require more cold feed bins. It is not uncommon to see up to six or seven cold bins used by some contractors, and those who do, normally use single size aggregates to minimize stockpile effects.

In summary, tall conical stockpiles of aggregates should be avoided because coarser particles tend to roll down the sides of the pile. Stockpiles constructed by layers or material inclines with less than 30 percent slopes are preferred. Ledging the stockpile is also an excellent way to minimize segregation at this stage. Constructing stockpiles on a hard foundation provides drainage, helps to reduce segregation, as well as contamination of the mineral aggregate, and is considered good practice

Figure 8.06 Building a Stockpile Properly

►►HMA Plant Production

Cold Bins Changes in aggregate supply and equipment require the loader operator to be more conscious of equipment operation. The manner in which the equipment enters the stockpile and operates can cause segregation in a stockpile or in a load of material. The loader operator should be alert for both segregation and variable moisture conditions.

Material placed in cold bins by a loader generally tends to form a conical shape, and coarse aggregates can separate at this point by rolling down the conical sides. On the other hand, the nature of the material draw onto the cold feed belt allows *reverse coning* to occur. Reverse coning is where coarse material falls to the middle of the cold bin, and it can lead to segregation on the belt and through the asphalt plant. One solution is to fractionate the aggregate; another is to keep the bins as full and level as possible. Yet another solution is to utilize self-relieving bottoms, which allow for a more uniform feeding of the cold aggregate all along the opening of the cold feed bin, eliminating bridging as a source of segregation. Cold bins are often equipped with a vibratory device to help provide a self-leveling effect.

Hot Bins Batch plant hot bins are prone to segregate by the screening process of the mineral aggregate. Screening action forces coarser aggregate to the far side of the bin and creates a pronounced pattern, from fine to coarse, at the inlet screen as shown in Figure 8.07. This pattern results from two factors: The way in which the aggregate flows across the screen; and the tendency for the finer material to slide down the near side of the bin. This process, if left uncorrected, will allow spotty segregation to occur. To correct this, baffle plates can be installed on the bin walls to remix material as it enters the bins. Even with the screening that they perform, batch plant hot bins do not fully correct segregation from poor stockpiling or cold feed operations.

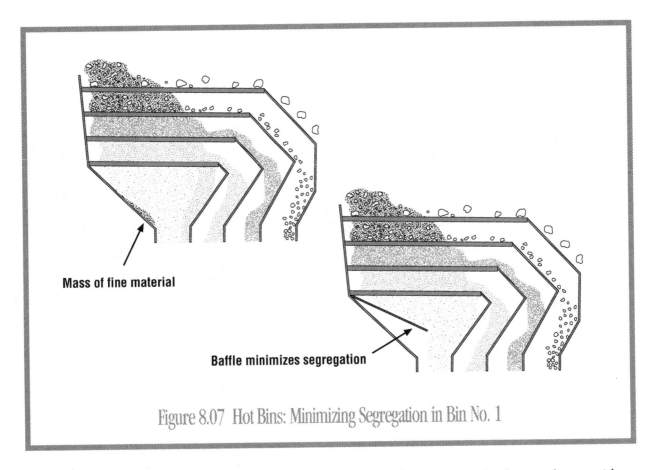

Mass of fine material

Baffle minimizes segregation

Figure 8.07 Hot Bins: Minimizing Segregation in Bin No. 1

Drum Mixer Operations In drum mix plants there are no hot bins. Stockpiles can be considered their equivalent, and there is no chance of segregated material remixing before the introduction of asphalt binder.

Segregation occurs in the drum, especially when starting and stopping production, due to the coarse material flowing through the drum at a slightly faster rate than the fine material. However, this can be alleviated by adjusting the starting and stopping times of the materials in the cold bins. Kickback flights or a dam can be installed in a drum to retard the flow and thereby increase the mixing time. In addition, the mixing time can be increased by extending the asphalt line farther into the drum or by decreasing the drum slope. An increase in the mixing time increases the asphalt coating of the aggregate and thereby the cohesion of the mix. This reduces the potential for segregation.

When the mix is discharged from the drum, the coarse and the fine particles tend to flow to opposite sides of the discharge chute. If this segregated material drops directly on a conveyor traveling on the line of the drum, the mix will remain segregated (Figure 8.08) all the way to the surge bin. Installation of a fixed plow in the drum (Figure 8.09), or otherwise restricting the discharge of the mix to a smaller area will decrease this segregation. Also, setting the drag conveyor at 90 degrees to the drum discharge (Figure 8.10) will eliminate or reduce segregation.

Hot Elevator Typically, asphalt mixes are discharged from the mixing unit and placed in surge or storage bins in two ways: drag slat and conveyor belt. To place material in the storage bin without segregation, the speed of these systems must be balanced with output tonnage. A small amount of material on a fast belt will segregate when it tumbles as it travels up the hot elevator, and when it is placed, or cast, into the bin. Coarser material will be cast to the far side of the drop chute or any other device installed to funnel material into the bin.

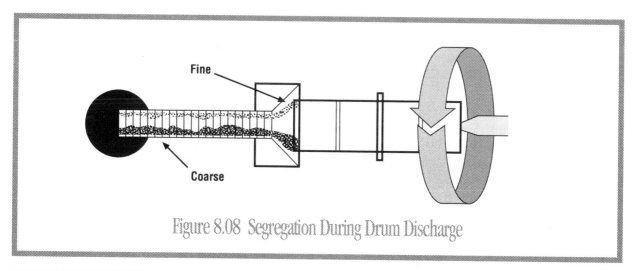

Figure 8.08 Segregation During Drum Discharge

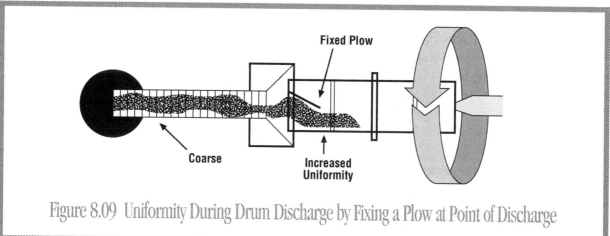

Figure 8.09 Uniformity During Drum Discharge by Fixing a Plow at Point of Discharge

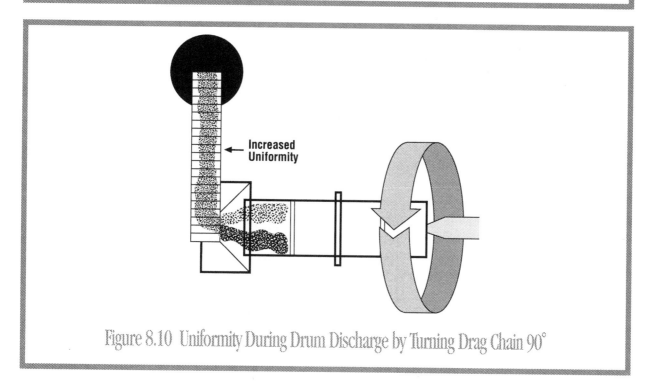

Figure 8.10 Uniformity During Drum Discharge by Turning Drag Chain 90°

Figure 8.11 Fine Aggregate Segregation of HMA on the Roadway

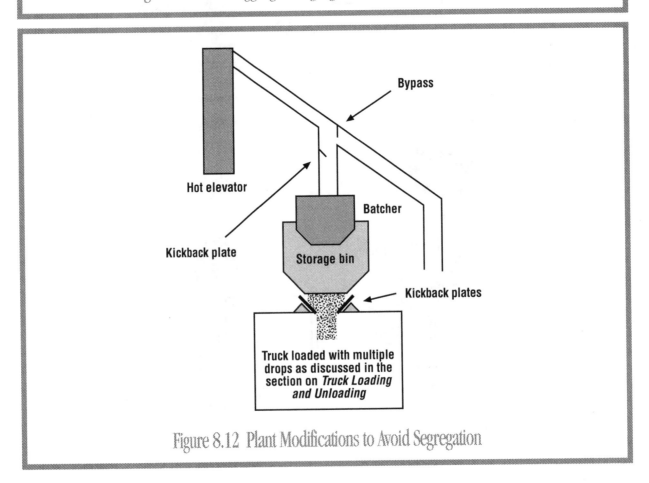

Figure 8.12 Plant Modifications to Avoid Segregation

Some storage systems allow material to be conveyed against a vertical side wall at the conveyor release point. This causes coarse aggregate to bounce back from the fine aggregate and produces a coarse-to-fine side throughout the bin. When dumped, the mix becomes coarse on one side of the truck and fine on the other, and results in continuous segregation along the side of a paving lane as shown in Figure 8.11.

Storage bins are normally fed through the use of a batcher, discussed later, and can be bypassed for direct loading if required. In these instances, segregation of the mixture bypassing the storage bin is very minimal. Segregation does occur when the bypass is locked off and the mixture now changes direction to enter the batcher. The installation of kickback plates, as shown in Figure 8.12, forces the remixing of coarse and fine aggregate within the mixture prior to entering the batcher. This technique is normally used on dated equipment, as most new units take this movement into account in their designs.

Pugmill and Bin Gates Discharge gate openings on pugmills are different from those on surge and storage bins. Batch plant pugmills empty all the mix in a single drop. Their gates open very wide and allow a straight drop into the truck. Trucks used in batch plant operations have traditionally been smaller and have therefore limited the coning effect in loading. Bin discharge gates that slide open slowly allow some dribble effect at the opening and closing mode. This is common to both the batch and drum mix plants that use surge or storage bins.

Several gate models are in use on surge and storage bins. They vary from rectangular to round, and can include plates and baffles. Gates are a common spot at which segregation occurs. Many agency specifications require clam or other type of gate openings that will not cause segregation. They open both transversely or longitudinally to the truck bed. Type of gate opening may be secondary to bin configuration in segregation control.

Surge and Storage Bins Segregation often occurs when material is placed into surge bins and storage bins, or "silos." A conveyor system places the mix into a "batcher," or similar discharge system. A batcher is a container with gates at the bottom, which sits above the silo and catches and releases material in order to regulate the flow of mix into the silo and reduce segregation. Timing systems on these units should be set in accordance with output tonnage to keep the mix from falling straight through the batcher into the silo and causing segregation by coning. For proper operation, the mix should be discharged into the silo in small batches. The batcher gate should be operated in a pulsating manner, and the batcher should never be completely empty during operation. For variable production rates, the gate timer should be adjusted accordingly.

Another method of reducing segregation in the silos is by the use of a rotating discharge chute at the top of the silo. In the rotating chute and plate systems, operation speed depends on tonnage output. Operating the device slowly will prevent material from being cast to the sides of the silo and causing a reverse cone effect, with coarse material on the outside and fine on the inside. The rotating chute has given way to the batcher in modern HMA plants.

Four silo variables govern the sensitivity of segregation:

- Diameter
- Height
- Cone shape
- Gate opening type

Each type of silo must be operated appropriately to obtain equal results. Diameter and height of the cone should be adequate to prevent reverse cone effects as mix leaves the unit. Segregation from these units can be corrected by adding another cone or batcher to the bottom. This remixes the material before it is released into trucks.

There are different types of silos with various side angles, gate openings, and inter-loading units. Some are more efficient than others. The ideal operating range for most silos is between 25 percent and 75 percent full. Segregation will increase below the 25 percent capacity as the mix falls below cone level.

Discharge Systems When segregation occurs in the silo, a cone unit placed on the bottom of the silo will produce reverse coning and remix the material. In addition, baffle plates can be installed on the bottom of the silo to kick coarse aggregate back into the mix. This can be seen in Figure 8.12, Plant Modifications to Avoid Segregation.

Retrofits to asphalt plants are a rather common occurrence. Preventing segregation is a good reason to look at retrofitting an asphalt plant. With some minor reconfiguration, benefits well beyond the cost of the retrofit can be obtained through increased pavement life.

Avoiding Segregation in Construction

➤➤ Trucking Loading and Unloading

HMA can segregate as it is loaded into trucks from either pugmill or surge systems. Trucks are much larger today, and care should be taken during loading to prevent coning in the truck bed. Both pug and surge systems are covered by the same loading recommendations. There should be at least three drops of material—front, back, and center, as shown in Figure 8.13. Even though it is more time consuming, proper care in truck loading can be extremely important to the quality of pavement produced.

The "length of slope" of a pile of HMA is of particular importance when considering truck loading. A single discharge into the truck tends to provide the longest slopes over which the stones can separate. The mix should discharge first near the bulkhead to produce a stacking effect for the front and sides. The second drop produces the same conditions at the rear of the truck. The third drop, in the middle, should overlap the conical sides of the first two. This procedure does not eliminate segregation, but spreads it out and minimizes the effect.

The method used to unload the haul truck into the paver hopper may also contribute to a segregation problem, again by generating longer slopes over which the particles may separate within the dumping load. Most problems of this kind develop from slowly raising the truck bed while dumping. When the truck is in place to dump onto the roadway, the bed should be partially elevated before the tailgate is released. This permits the mix to move in mass and to flood the hopper, thus preventing the coarse aggregate from falling out first and causing spotty segregation patterns.

➤➤ Paver Operation

In addition to the operations through the plant and the truck loading, the possibility of segregation in the paving operation exists. Segregation that is intermittent can usually be related to the trucks or the loading procedure. Continuous segregation can generally be attributed to operations and material handling occurring in the paver. An intermittent type of segregation that is an exception and shows up on the road as an open textured "chevron," pointing in a direction opposite to that being paved, is illustrated in Figure 8.14. In most cases the spacing of these open textured areas is end-of-load associated. Dumping the wings, thus sending coarser material to the middle of the hopper, after running the paver dry and then moving forward before the hopper is re-charged can create this form of segregation.

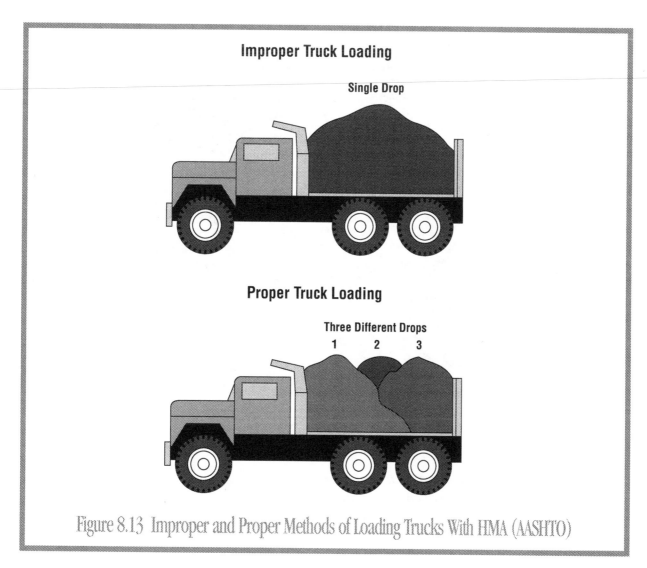

Improper Truck Loading

Single Drop

Proper Truck Loading

Three Different Drops
1   2   3

Figure 8.13 Improper and Proper Methods of Loading Trucks With HMA (AASHTO)

If the hopper wings are dumped, care should be taken when dumping them, and it should only be done with other, non-segregated material in the feeder area. While this may not completely solve the problem, it will give the coarse material an opportunity to mingle with more homogeneous material, thus minimizing the segregation effect. Dumping the wings and running the hopper dry before recharging causes chevron or wing type segregation. Once the paver is in motion, the hopper should maintain 25 percent or more of its capacity at all times. The paver should not be operated when the material is below the hopper deck.

►►*Paver Segregation Areas*

Both continuous and repeated forms of segregation can be linked to the paver and, in particular, to its material handling system. There are specific areas across the mat where visible segregation may occur. These are at the:

- Center screw conveyor (auger) support
- Outside edges of the flight feeders (drag conveyors)
- Underside of the outboard screw conveyor (auger) supports
- Outer edges of the screed

Figure 8.14 Chevron-Shaped Segregation on the Roadway

Center Screw Conveyor Support, or Center Gear Box A longitudinal, segregated strip often develops in the center of the paved lane under the center screw conveyor support. This segregation is not always visible at the time of paving. More recent developments in screw conveyor (auger) systems that can be raised and lowered have shown reasonable success in helping solve the problem. This new approach utilizes a separate drive motor and case for the screw conveyor drive and eliminates the previous triangular-shaped drive case. In this way, feeder discharge can directly surge fill the cavity under the screw conveyor drive case. Success has also been achieved in eliminating the segregated area by simply raising the screw conveyors to allow unrestricted material flow into the center area. Once the cavity is filled, the roll distances are shortened and segregation is minimized, but operator care must be taken to ensure that a proper screw conveyor chamber level is maintained between truckloads to keep the cavity filled.

Outside Edges of the Flight Feeders A few mixes have shown a tendency to segregate at the outside edges of the flight feeders for the same "fall distance" reasons as the center support. The mix could also segregate from flowing around the corner and then back under the paver's main frame. Most new pavers are equipped with a baffle to minimize the flight feeder conveyors' ability to carry material forward on their return path. This baffle has also been successful in reducing the roll distance at that point.

The most important action to reduce segregation in these areas is to keep the screw conveyor chamber filled to a level at or about the top of the conveyor shaft. This fills the cavity ahead of the flight feeders and minimizes the opportunity for the material to flow forward after turning the corner.

Under the Outboard Screw Conveyor Supports As the HMA moves outward axially along the screw conveyor (auger), it reaches the outboard conveyor support and bearing, which can develop a resistance to continued flow. Beyond that point, movement of the mix is achieved by pushing it outward using the influence of the last inboard screw flight's rotation.

Generally, problems rarely develop in this area with most easy-flowing mixes. However, stiffer, flow resistant mixes can cause a problem by gathering ahead of the screw bearing and flowing ahead under the paver's main frame, with some risk of mix segregation. In these cases, it is wise to use a short screw conveyor segment outside of the support and bearing to control and "power" the mix outside and away from the bearing. This will minimize the potential build-up and problems. The rule should be not to add the segment automatically unless it is necessary

Outer Edges of the Screed The most successful solution to eliminating segregation at the outer edges of the screed is maintaining a proper head of material at the screed's end gates (edger plates). This should be the same level as the rest of the conveyor—right at or about the top of the conveyor shaft. It is probably better to fault to a higher rather than lower head of material in this case, the key being to keep the level consistent. Continuous forward movement is important in keeping this balance.

➤➤ *Function of Feeder (Flow) Gates*

With more modern pavers featuring self-regulating, proportional flight feeder/screw conveyor drives, the role of the feeder (flow) gates in regulating the level of material in the screw conveyor chamber is not always understood at the project level. These gates play a very important role in establishing the different capacities between the feeder and the screw conveyor, and they do not control total feeder/screw volume as was formerly experienced with older machines that had "on-off" feeder systems.

One of the more important tasks in minimizing segregation is to keep the proper head of material in the screw conveyor chamber. It is important to understand how the function and adjustment of the feeder gates contributes to maintaining that level. Their function can probably be best described by the following example.

Assume that a single flight feeder is required to supply 150 tons per hour (TPH) to meet the demand of the screed for that half pavement width. Required capacity is generated by the speed of the flight feeder in combination with the open area under the feeder gate. It must be understood that the 150 TPH requirement can be provided at a lower speed and greater area (wider gate opening) or a higher speed and lesser area (lower gate opening). In either case, the demand for material is established by the combination of the screed width, layer thickness, and paving speed. The sensing device will establish the speed of the feeder to meet that demand, regardless of the gate opening, as long as that requirement is within the maximum production capacity of the feeder.

The screw conveyor is linked to the flight feeder either by a roller chain or hydraulically. Its capacity, then, is the result of a combination of its shaft speed and the volume of material established by the area difference generated between its outer and inner diameters and the distance between its flights (its pitch).

If a shortage of material is noticed at the edges of the screed (usually also noticed as too much material at the center of the screed), the screw conveyor's capacity must be increased to move more material to the outside. In this case, the feeder gates must be lowered. This will reduce the area under the gate and cause the feeder to speed up in order to continue to meet the 150 TPH demand. In so doing, however, the mechanical or hydraulic linkage also causes the screw conveyor to speed up, thus increasing its relative capacity and providing more material to the end of the screed.

Similarly, too much material at the screed ends requires the gates to be raised to reduce the feeder/screw speed and diminish the screw conveyors' ability to move material to the screed edge.

►► Screed and Screw Conveyor Extensions

Self-extending screeds with trailing extenders are, by design, naturally self-feeding with most mixes and can generally be operated without screw conveyor extensions. A natural material flow is created as the screed is towed into the HMA supply, which is at rest on the paving surface. As the inside corner of the transition between the main screed and the trailing extender becomes filled with material, the natural action of adding more material in that area causes the mix to move outward to the screed's extremities, generating a "waterfall" effect.

Sometimes, an unnecessarily extended screw conveyor passing through this natural repose can only promote the risk of mix segregation as the screw flight rotates, particularly if it is operated at extra high speed. In these cases, screw conveyor extensions should only be used if the flow characteristics of the mix do not allow free flow, and the mix requires additional force and control to move it outward properly. The rule should be based on an end result criterion stating that "the paver material handling system should provide non-segregated material to the end-gate (edger plate) in sufficient quantities to provide for the variable pavement thickness requirements expected for that particular job." Screw conveyor extensions should only be required with self-extending screeds when such an end-result directed criterion cannot be met.

Analysis of Segregation

►► Pinpointing Segregation

All mixes and test samples, by nature, are subject to a certain degree of variation from the job mix formula. Limits on these variations, established by the agency based on its experience with mixes and sampling procedures, vary somewhat. However, when analyzing a segregated HMA mix, if a substantial amount of data is available, the agency can determine patterns in the mixture composition. Asphalt content and gradation of samples taken at strategic locations will furnish the information needed to pinpoint where segregation is occurring and to what extent.

To accomplish this, the results of asphalt content and aggregate gradation tests on a number of samples of HMA are used. Typically, the percent of material passing a selected critical sieve, usually the 2.36 mm (No. 8) or 2.00 mm (No. 10) sieve, is plotted on a graph against the asphalt content of that sample. Test results should be arranged according to the source of the samples, therefore, test results from samples taken at the HMA plant would be plotted separately from those originating at the roadway. This will facilitate interpretation of the test data and provide better information on where segregation occurs.

The graph indicates the percents passing the critical sieve on the horizontal axis, arranged in ascending order from the origin. The asphalt contents are arranged in ascending order from the origin on the vertical axis. The analysis is made using the asphalt content for each extraction

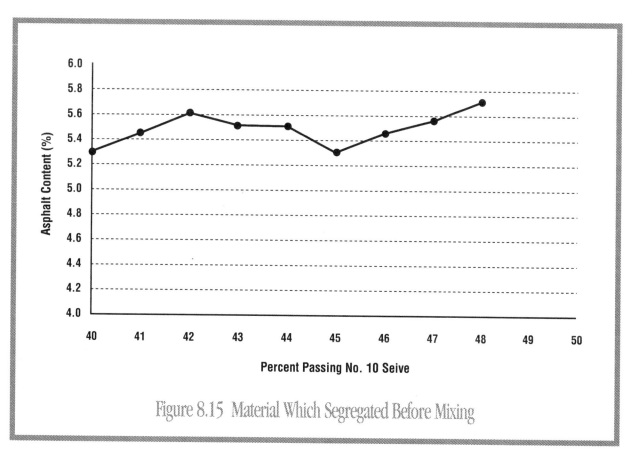

Figure 8.15 Material Which Segregated Before Mixing

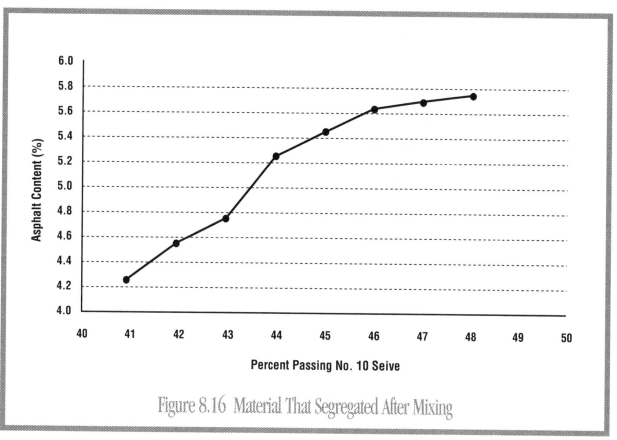

Figure 8.16 Material That Segregated After Mixing

plotted against its percentage passing the critical sieve. This procedure is followed for all of the samples from each selected location. The data is plotted as shown in Figure 8.15.

If the plot of these points from samples of segregated mix is roughly horizontal or nonlinear, the aggregate probably segregated before mixing as shown in Figure 8.15. This plot indicates that, for varying percents of material passing the 2.00 mm (No. 10) sieve, the amount of asphalt in the mix is erratic and has no trend, either upward or downward. To find the possible source of segregation in this case, the aggregate source, aggregate handling operation, cold bin loading and feeding, and drum or batch mixing operations should be checked.

If the plot of these points results in a relatively straight line, increasing in asphalt content as the percent passing the 2.00 mm (N0. 10) sieve increases, the HMA probably segregated after mixing. This is illustrated in Figure 8.16. It is normal for finer mixes to have a higher asphalt content. The ascending plot indicates that the finer particles got approximately the same coating of asphalt as they were being mixed, so that when the percent passing the 2.00 mm (No. 10) sieve increases, the asphalt content of the mix does also. If the finer particles got about the same coating, the proportion of coarse and fine material in the mixing facility was probably constant. In other words, the material was not segregated at that time. The places to look for causes of segregation in this case are the hot elevator, mixture storage and discharge facilities, truck loading operations, and paver operation.

Many of the poor practices occurring can be caught through a quick review of material handling of the finished product. However, sometimes a more thorough review of plant operations will be required to pinpoint the actual cause of segregation at this phase. If mineral aggregate handling and mixture handling procedures appear adequate, then this is a good time to review the mixture design and take appropriate action by perhaps modifying the job mix formula.

When the extraction of a base or intermediate course involves larger aggregate, the sample may be specially prepared to insure a representative sample. This procedure involves taking a large sample and separating it into plus 12.5 mm and minus 12.5 mm (+0.5 inch and -0.5 inch) portions. A ratio of these portions to the total sample is used to proportion the weight of the plus 12.5 mm and minus 12.5 mm material. Proportionate amounts of quartered samples are remixed for the extraction test, and this procedure assures the test sample will represent the mix.

►►Determining Segregation Level

Guidelines currently being used to identify segregation typically follow a standard format. It appears that most have been developed jointly by agency and industry in order to achieve the common goal. Methods of identification typically used are:

- Conducting preliminary field evaluations of existing projects exhibiting segregation, in order to determine and visually classify various levels of segregation.
- Developing a method to evaluate unacceptable segregation on construction projects.
- Verifying the test method on an active construction project.
- Developing an end product limit for segregation.

These field evaluations must be completed in order to clearly define achievable goals for all parties concerned. First, the evaluations give both parties the opportunity to walk the pavement together to see what a certain level of segregation looks like and to allow the agency to define what is an unacceptable level requiring corrective action. Second, the evaluations will allow the parties to determine if the various test methods used to measure segregation, measure the varying levels of segregation accurately. All written standards may include the following as minimums.

- Description of segregation
- Definitions of types and severity of segregation
- QC responsibilities of the contractor regarding segregation
- Investigations
- Dispute resolutions
- Acceptable corrective actions
- Basis of payment

Minimizing the likelihood of placing segregated mixture is the intent of all segregation standards. Remedial work, removal and replacement, and penalties are all methods of corrective action being enforced. However, when the pavement is deemed defective with regard to segregation, on the basis of visual inspection, the issue should be brought to the contractor's attention, in writing, as soon as possible. The actual severity and extent of segregation is typically reported at a later date. Understanding that severity levels exist for segregation is of paramount importance. Severity of segregation can be categorized as follows:

- Slight: Area where the mastic is in place between aggregate particles; however, there is slightly more coarse aggregate than in the surrounding acceptable mix.
- Medium: Area has significantly more coarse aggregate than the surrounding acceptable mat and usually exhibits some lack of surface mastic.
- Severe: Area appears very coarse in comparison to the surrounding acceptable mat, with stone against stone, and little or no mastic.

When severe segregation in any pavement course, or medium segregation in a surface course, is identified, a segregation standard should require the contractor to propose in writing the corrective action to be taken. Corrective actions vary between jobs and mixtures and will usually require actions similar to those described in earlier sections. Agencies normally review the corrective work proposal and consider the following actions to be acceptable for the types of segregation defined above.

- Slight: Typically will be accepted into the work.
- Medium: Generally left in place for lower layers; however, surface courses are typically subject to price adjustments, removal and replacement, or resurfacing, at the contractor's cost.
- Severe: Often the pavement is to be removed and replaced across the full lane or shoulder width in a workmanlike manner.

Work is typically suspended when the medium or severe level of segregation is identified. This allows the contractor the opportunity to address and take corrective action against any further segregation on the current project. Written plans, which could become contract documents, are usually submitted by the contractor to the agency before continuing paving operations.

Quantifying levels of segregation has become an issue in combating segregation. Many agencies use the following for quantifying segregation.

- Deviation from the approved job mix formula (JMF)
- Sand patch measurement
- Nuclear density gauge measurements
- Visual observation of non-uniform texture

Typical deviations allowed from approved JMF's for aggregate gradations are well established. When an extraction indicates a deviation greater than the established limit, segregation probably exists. The sand patch test (ASTM E965), when performed properly, demonstrates a mixture's ability to hold water. In this instance, the water is replaced with sand, which should cover a certain area when broomed around. The area covered will be less when placed over a severely segregated area. Some agencies use a nuclear density gauge to quantify segregation as follows:

After rolling, three density readings are made at a selected transverse location at specified longitudinal distances from the truck exchange point. Excessively high or low values are thrown out, and the remaining values are averaged. Differences between average value and longitudinal values are calculated and recorded. Differences above maximum allowable are cause for shutdown because of segregation. Agencies may specify a maximum difference of 80 kilograms per cubic meter (5 lbs. per cubic foot) on base or binder course, and a maximum difference of 64 kilograms per cubic meter (4 lbs. per cubic foot) on a surface course as indicators for identifying these areas.

Given the financial impact to an agency's maintenance costs, these guidelines appropriately continue to be written and included in contract documents. Uniform and committed enforcement of segregation guidelines can substantially decrease placement of defective HMA pavements. Segregation is a defect that must be reduced to increase the life of HMA pavements.

Table A.01 Conversion Factors (Customary to Metric Units)

To convert from	To	Multiply by
acre	meter^2 (m^2)	4046.856
acre	hectometer^2 (hm^2)	0.404686
Atmosphere (technical = 1kgf/cm^2)	kilopascal (kPa)	98.0665
barrel (42 gal.)	decimeter^3 (dm^3) or liter (l)	158.987.3
BTU (international Table)	kilojoule (kJ)	1.055056
bushel	decimeter^3 (dm^3)	35.2391
dyne	micronewton (μN)	10.0
dyne/centimeter^2	pascal (Pa)	0.1
Fahrenheit (temperature)	Celsius (°C)	$t_c = (t_f-32)/1.8$
foot	meter (m)	0.3048
foot^2	meter^2 (m^2)	0.092903
foot^3	meter^3 (m^3)	0.028317
	liter (l)	28.317
foot-pound-force	joule (j)	1.355818
foot/minute	meter/second (m/s)	0.00508
foot/second	meter/second^2 (m/s^2)	0.3048
gallon (U.S. liquid)	decimeter^3 (dm^3) or liter (l)	3.785412
	meter^3 (m^3)	0.003785
gallon/minute	decimeter^3/second (dm^3/s) or liter/second (l/s)	0.06309
gallon/yard^2	decimeter^3/meter^2 (dm^3/m^2) or liter/meter^2 (l/m^2)	4.527314
horsepower (electric)	kilowatt (kW)	0.746
inch	millimeter (mm)	25.4
inch^2	centimeter^2 (cm^2)	6.4516
inch^2	millimeter^2 (mm^2)	645.16
inch^3	centimeter^3 (cm^3)	16.38706
inch/second	meter/second (m/s)	0.0254
inch of mercury (60? F)	pascal (Pa)	3376.85
inch/second^2	meter/second^2 (m/s^2)	0.0254
kilogram (kg)	ton (metric	0.001
kip (1000 lbf)	kilonewton (kN)	4.448222
kip/inch^2	megapascal (MPa)	6.894757
mile (U.S. statute)	kilometer (km)	1.609344
mile^2	kilometer^2 (km^2)	2.589988
mile/hour	kilometer/hour (km/hr)	1.609344
minute (angle)	radian (rad)	0.00029089
ounce-force	newton (N)	0.278 0139
ounce-mass	gram (g)	28.34952
ounce-fluid	centimeter^3 (cm^3)	29.57353
	liter(l)	0.029574
pint (U.S. liquid)	liter (l)	0.4731765
poise (absolute viscosity)	pascal-second (Pa-s)	0.1
pound-force (lbf)	newton (N)	4.448222
	kilonewton (kN)	0.004448
pound-force-inch	newton-meter (N.m)	0.1129848
pound-force/foot^2	pascal (Pa)	47.88026
pound-force/inch^2 (psi)	kilopascal (kPa)	6.894757
pound-mass	kilogram (kg)	0.4535924
pouound-mass/foot^2	kilogram/meter^2 (kg/m^2)	4.882428
pound-mass/foot^3	kilogram/meter^3 (kg/m^3)	16.01846
	megagram/meter^3(Mg/m^2)	0.-016018
pound-mass/inch^3	kilogram/decimeter^3 (kg/dm^3)	27.67990
pound-mass/gallon (U.S. liquid)	kilogram/meter^3 (kg/dm^3)	119.8264
	kilogram/decimeter^3 (kg/dm^3)	0.119.826
psi	kilopascal (kPa)	6.894757
quart (U.S. liquid)	decimeter^3 (dm^3) or liter (l)	0.9463529
ton (metric)	kilogram (kg)	1000.0
ton (short-2000 lb)	kilogram (kg)	907.1847
ton (long-2240 lb)	kilogram (kg)	1016.0461
ton-mass/yard^3	kilogram/meter^3 (kg/m^3)	1186.5527
yard	meter (m)	0.9144
yard^2	meter^2 (m^2)	0.8361274
yard^3	meter^3 (m^3)	0.7645549

Table A.02 Weight and Volume Relations for Various Types Of Compacted Asphalt Pavements

kg/m³	kg/dm³	kg/m²/cm depth	lb/ft³	lb/yd³	lb/yd²/in depth
1600	1.6	16	100	2700	75
1700	1.7	17	105	2835	79
1800	1.8	18	110	2970	82
1900	1.9	19	115	3105	86
2000	2.0	20	120	3240	90
2100	2.1	21	125	3375	94
2200	2.2	22	130	3510	97
2300	2.3	23	135	3645	101
2400	2.4	24	140	3780	105
2500	2.5	25	145	3915	109
2600	2.6	26	150	4050	112
			155	4185	116
			160	4320	120

Range for Various Pavement Types

Pavement Type	kg/m³	kg/dm³	kg/m²/cm depth	lb/ft³	lb/yd³	lb/yd²/in depth
Open Graded	1850-2250	1.85-2.25	18.5-22.5	115-140	3105-3780	86-105
Coarse Graded	2100-2400	2.10-2.40	21.0-24.0	130-150	3510-4050	97-112
Dense Graded	2150-2500	2.15-2.50	21.5-25.0	135-155	3645-4185	101-116
Fine Graded	2100-2400	2.10-2.40	21.0-24.0	130-150	3510-4050	97-112
Stone Sheet	2100-2400	2.10-2.40	21.0-24.0	130-150	3510-4050	97-112
Sand Sheet	1900-2250	1.90-2.25	19.0-22.5	120-140	3240-3780	90-105
Fine Sheet	1900-2250	1.90-2.25	19.0-22.5	120-140	3240-3780	90-105
Mixed-in-Place Macadam	1750-2150	1.75-2.15	17.5-21.5	110-135	2970-3645	82-101
Mixed-in-Place Dense Graded	1750-2150	1.75-2.15	17.5-21.5	110-135	2970-3645	82-101
Mixed-in-Place Sand Asphalt	1600-2000	1.60-2.00	16.0-20.0	100-125	2700-3375	75-94

Note:
Because of the considerable variations in specific gravity, gradation, and other characteristics of mineral aggregates, weight per unit volume of compacted asphalt pavement varies considerably. Exact weights per unit volume should be determined in the laboratory from samples taken from the same materials as used in the field.

Preliminary Estimating

kg/m²/cm depth	lb/yd²/in depth
21.0	100
22.5	105
23.5	110
22.5	105
22.5	105
21.0	100
21.0	100
20.0	95
20.0	95
18.0	95

Table A.03 Composition of Asphalt Concrete

Sieve Size	Mix Designation and Nominal Maximum Size of Aggregate				
	37.5 mm (1-1/2 in.)	25.0 mm (1 in.)	19.0 mm (3/4 in.)	12.5 mm (1/2 in.)	9.5 mm (3/8 in.)
	Total Percent Passing (by weight)				
50.0 mm (2 in.)	100	—	—	—	—
37.5 mm (1-1/2 in)	90 to 100	100	—	—	—
25.0 mm (I in.)	—	90 to 100	100	—	—
19.0 mm (3/4 in.)	56 to 80	—	90 to 100	100	—
12.5 mm (1/2 in.)	—	56 to 80	—	90 to 100	100
9.5 mm (3/8 in.)	—	—	56 to 80	—	90 to 100
4.75 mm (No. 4)	23 to 53	29 to 59	35 to 65	44 to 74	55 to 85
2.36 mm (No. 8)*	15 to 41	19 to 45	23 to 49	28 to 58	32 to 67
1.18 mm (No. 16)	—	—	—	—	—
0.60 mm (No. 30)	—	—	—	—	—
0.30 mm (No. 50)	4 to 16	5 to 17	5 to 19	5 to 21	7 to 23
0.15 mm (No. 100)	—	—	—	—	—
0.075 mm (No. 200)**	0 to 5	1 to 7	2 to 8	2 to 10	2 to 10
Asphalt Cement, weight percent of Total Mixture†	4 to 11	3 to 8	3 to 9	4 to 10	5 to 12

*In considering the total grading characteristics of an asphalt paving mixture, the amount passing the 2.36 mm (No. 8) sieve is a significant and convenient field control point between fine and coarse aggregate. Gradings approaching the maximum amount permitted to pass the 2.36 mm (No. 8) sieve will result in pavement surfaces having comparatively fine texture, while gradings approaching the minimum amount passing the 2.367 mm (No. 8) sieve will result in surfaces with comparatively coarse texture.

**The material passing the 0.075 mm (No. 200) sieve may consist of fine particles of the aggregates or mineral filler or both. It shall be free from organic matter and clay particles and have a plasticity index not greater than 4 when tested in accordance with Method T 89 and Method T 90.

†The quantity of asphalt cement is given in terms of weight percent of the total mixture. The wide difference in the specific gravity of various aggregates, as well as a considerable difference in absorption, results in a comparatively wide range in the limiting amount of asphalt cement specified. The amount of asphalt required for a given mixture should be determined by appropriate laboratory testing or the basis of past experience with similar mixtures or by a combination of both.

Aggregate: A hard inert material of mineral composition such as sand, gravel, slag, or crushed stone, used in pavement applications either by itself or for mixing with asphalt.

Types:

Coarse Aggregate: Aggregate retained on the 2.36 mm (No. 8) sieve.

Coarse-Graded Aggregate: One having a continuous grading in sizes of particles from coarse through fine with a predominance of coarse sizes.

Dense-Graded Aggregate: An aggregate that has a particle size distribution such that when it is compacted, the resulting voids between the aggregate particles, expressed as a percentage of the total space occupied by the material, are relatively small.

Fine Aggregate: That passing the 2.36 mm (No. 8) sieve.

Fine-Graded Aggregate: One having a continuous grading in sizes of particles from coarse through fine with a predominance of fine sizes.

Open-Graded Aggregate: One containing little or no mineral filler and in which the void space in the compacted aggregate are relatively large.

Well-Graded Aggregate : Aggregate graded with relatively uniform proportions from the maximum size down to filler with the object of obtaining an asphalt mix with a controlled void content and high stability.

Aggregate Storage Bins*:   Bins that store the necessary aggregate sizes and feed them to the dryer in substantially the same proportions as are required in the finished mix.

Air Voids: Empty spaces in a compacted mix surrounded by asphalt-coated particles, expressed as a percentage by volume of the total compacted mix.

Asphalt*: A dark brown to black cementitious material in which the predominating constituents are bitumens which occur in nature or are obtained in petroleum processing. Asphalt is a constituent in varying proportions of most crude petroleums.

Asphalt Binder: Asphalt cement that is classified according to the Standard Specification for Performance Graded Asphalt Binder, AASHTO Designation MP 1. It can be either unmodified or modified asphalt cement, as long as it complies with the specifications.

Asphalt Cements: A fluxed or unfluxed asphalt specially prepared as to quality and consistency for direct use in the manufacture of asphalt pavements.

Asphalt Concrete: See Hot Mix Asphalt.

Asphalt Leveling Course: A course of hot mix asphalt of uniform or variable thickness used to eliminate irregularities in the contour of an existing surface prior to placing the subsequent course.

Asphaltenes*: The high molecular weight hydrocarbon fraction precipitated from asphalt by a designated paraffinic naphtha solvent at a specified solvent-asphalt ratio.

Automatic Cycling Control*: A control system in which the opening and closing of the weigh hopper discharge gate, the bituminous discharge valve, and the pugmill discharge gate are actuated by means of self-acting mechanical or electrical machinery without any intermediate manual control. The system includes preset timing devices to control the desired periods of dry and wet mixing cycles.

* ASTM Definitions

APPENDIX B

GLOSSARY OF TERMS PERTAINING TO ASPHALT PAVEMENT CONSTRUCTION

B-1

Automatic Dryer Control*: A system that automatically maintains the temperature of aggregates discharged from the dryer within a preset range.

Automatic Proportioning Control*: A system in which proportions of the aggregate and asphalt fractions are controlled by means of gates or valves, which are opened and closed by means of self-acting mechanical or electronic machinery without any intermediate manual control.

Bank Gravel*: Gravel found in natural deposits, usually intermixed with fine material such as sand or clay or combinations thereof; includes gravelly clay, gravelly sand, clayey gravel, and sandy gravel (the names indicate the relative proportions of the materials in the mixture).

Base Course: The layer of material immediately beneath the surface or intermediate course. It may be composed of crushed stone, crushed slag, crushed or uncrushed gravel and sand, or of hot mix asphalt, typically with larger size aggregate.

Batch Plant*: A manufacturing facility for producing asphalt paving mixtures that proportions the aggregate constituents into the mix by weighed batches and adds asphalt material by either weight or volume.

Bitumen*: A class of black or dark-colored (solid, semisolid, or viscous) cementitious substances, natural or manufactured, composed principally of high molecular weight hydrocarbons, of which asphalts, tars, pitches, and asphaltites are typical.

Blast-Furnace Slag*: The nonmetallic product, consisting essentially of silicates and alumino-silicates of lime and of other bases, that is developed simultaneously with iron in a blast furnace.

Bleeding or Flushing: The upward movement of asphalt in an asphalt pavement resulting in the formation of a film of asphalt on the surface. The most common cause is too much asphalt in one or more of the pavement courses, resulting from too rich a plant mix, an improperly constructed seal coat, too heavy a prime or tack coat, or solvent carrying asphalt to the surface. Bleeding or flushing usually occurs in hot weather.

Clinker*: Generally a fused or partly fused by-product of the combustion of coal. Also includes lava and portland-cement clinker and partly vitrified slag and brick.

Coal Tar*: A dark brown to black cementitious material produced by the destructive distillation of bituminous coal.

Compaction: The act of compressing a given volume of material into a smaller volume. Insufficient compaction of the asphalt pavement courses may accelerate the onset of pavement distresses of various types.

Consensus Properties: Aggregate characteristics that are critical to well-performing hot mix asphalt, regardless of the aggregate source, and whose limiting values are set by the Superpave specification.

Consistency: The degree of fluidity of asphalt cement at any particular temperature. The consistency of asphalt cement varies with its temperature; therefore, it is necessary to use a common or standard temperature when comparing the consistency of one asphalt cement with another.

Corrugations (Washboarding) and Shoving: Types of pavement distortion. Corrugation is a form of plastic movement typified by ripples across the asphalt pavement surface. Shoving is a form of plastic movement resulting in localized bulging of the pavement surface. These distortions usually occur at a points where traffic starts and stops, on hills where vehicles brake on the downgrade, on sharp curves, or where vehicles hit a bump and bounce up and down. They occur in asphalt layers that lack stability. Lack of stability may be caused by a mixture that is too rich in asphalt, has too high a proportion of fine aggregate, has coarse or fine aggregate that is too round or too smooth, or has asphalt cement that is too soft. It may also be due to excessive moisture and/or contamination due to oil spillage.

Cracks: Breaks in the surface of an asphalt pavement. The common types are:

Alligator Cracks: Interconnected cracks forming a series of small blocks resembling an alligator's skin or chicken-wire, and caused by excessive deflection of the surface over unstable subgrade or lower courses of the pavement.

Edge Joint Cracks: The separation of the joint between the pavement and the shoulder, commonly caused by the alternate wetting and drying beneath the shoulder surface. Other causes are shoulder settlement, mix shrinkage, and trucks straddling the joint.

Lane Joint Cracks: Longitudinal separations along the seam between two paving lanes caused by a weak seam between adjoining spreads in the courses of the pavement.

Reflection Cracks: Cracks in asphalt overlays that reflect the crack pattern in the pavement structure underneath. They are caused by vertical or horizontal movements in the pavement beneath the overlay and brought on by expansion and contraction with temperature or moisture changes.

Shrinkage Cracks: Interconnected cracks forming a series of large blocks, usually with sharp corners or angles. Frequently they are caused by volume change in either the asphalt mix or in the base or subgrade.

Slippage Cracks: Crescent-shaped cracks that are open in the direction of the thrust of wheels on the pavement surface. They result when a severe or repeated shear stresses are applied to the surface and there is a lack of good bond between the surface layer and the course beneath.

Crusher-Run*: The total unscreened product of a stone crusher.

Cutback Asphalt: Asphalt cement which has been liquefied by blending with a petroleum solvent (also called a diluent), to form one of the following cutback asphalts. Upon exposure to atmospheric conditions the diluents evaporate, leaving the asphalt cement to perform its function.

Rapid-Curing (RC) Asphalt: Cutback asphalt composed of asphalt cement and a naphtha or gasoline-type diluent of high volatility.

Medium-Curing (MC) Asphalt: Cutback asphalt composed of asphalt cement and a kerosene-type diluent of medium volatility.

Slow-Curing (SC) Asphalt: Cutback asphalt composed of asphalt cement and oils of low volatility.

Road Oil: A heavy petroleum oil, usually similar to one of the slow-curing (SC) grades.

Delivery Tolerances*: Permissible variations from the exact desired proportions of aggregate and bituminous material as delivered into the pugmill.

Density: The degree of solidity that can be achieved in a given mixture, which will be limited only by the total elimination of voids between particles in the mass.

Densification: The act of increasing the density of a mixture during the compaction process.

Distortion: Any change of the pavement surface from its original shape.

Drum Mix Plant: A manufacturing facility for producing asphalt paving mixtures that proportions the aggregate, then dries and mixes the aggregate with a proportional amount of asphalt in the same drum. Variations of this type of plant use several types of drum modifications, separate (and smaller) mixing drums, and coating units (coater) to accomplish the mixing process.

Dryer*: An apparatus that will dry the aggregates and heat them to the specified temperatures.

Ductility: The ability of a substance to be drawn out or stretched thin. While ductility is considered and important characteristic of asphalt cements in many applications, the presence or absence of ductility is usually considered more significant that the actual degree of ductility.

Durability: The property of an asphalt paving mixture that represents its ability to resist disintegration by weathering and traffic. Included under weathering are changes in the characteristics of the asphalt, such as oxidation and volatilization, and changes in the pavement and aggregate due to the action of water, including freezing and thawing.

Emulsified Asphalt: A combination of asphalt cement, water and a small amount of an emulsifying agent. It is a heterogeneous system (containing two normally immiscible substanous phases: asphalt and water), in which the water forms the continuous phase of the emulsion, and minute globules of asphalt form the discontinuous phase. Emulsified asphalt may be either anionic – electronegatively charged asphalt globules – or cationic – electropositively charged asphalt globules – depending upon the emulsifying agent.

Emulsified Asphalt Mix (Cold Mix): A mixture of emulsified asphalt and aggregate produced in a central plant (plant mix) or mixed at the road site (mixed-in-place).

Fatigue Resistance: The ability of asphalt pavement to withstand slight repeated flexing, or bending, caused by the passage of wheel loads. As a rule, the higher the asphalt content, the greater the fatigue resistance.

Flexibility: The ability of an asphalt pavement structure to conform to settlement of the foundation. Generally, flexibility of the asphalt paving mixture is enhanced by high asphalt content.

Full-Depth® Asphalt Pavement: The term Full-Depth (registered by The Asphalt Institute with the U.S. Patent Office) certifies that the pavement is one in which asphalt mixtures are employed for all courses above the prepared subgrade.

Grade Depressions: Localized low areas of limited size, which may or may not be accompanied by cracking.

Hot Aggregate Storage Bins*: Bins that store heated and separated aggregates prior to their final proportioning into the mixer.

Hot Mix Asphalt (Asphalt Concrete): High quality, thoroughly controlled hot mixture of asphalt binder (cement) and well-graded, high quality aggregate, which can be thoroughly compacted into a uniformly dense mass.

Impermeability: The resistance an asphalt pavement has to the passage of air and water into or through the pavement.

Lift: A layer or course of paving material applied to a base or a previous layer.

Mesh*: The square opening of a sieve.

Mineral Dust: The portion of the fine aggregate passing the 0.075 mm (No. 200) sieve.

Mineral Filler: A finely divided mineral product, at least 70 percent of which will pass a 0.075 mm (No. 200) sieve. Pulverized limestone is the most commonly manufactured filler, although other stone dust, hydrated lime, portland cement, and certain natural deposits of finely divided mineral matter are also used.

Natural (Native) Asphalt: Asphalt occurring in nature, which has been derived from petroleum through natural processes of evaporation of volatile fractions, leaving the asphalt fractions. The native asphalt of most importance is found in the Trinidad and Bermudez Lake deposits. Asphalt from these sources is often called lake asphalt. Practically none of this asphalt is used in the United States today.

Open-Grade Asphalt Friction Course: A pavement surface course that consists of a high-void, asphalt plant mix that permits rapid drainage of rainwater through the course and out the shoulder. The mixture is characterized by a large percentage of one-sized coarse aggregate. This course prevents tire hydroplaning and provides a skid-resistant pavement surface.

Pavement Structure: A pavement, including all of its courses of asphalt-aggregate mixtures, or a combination of asphalt courses and untreated aggregate courses, placed above the subgrade or improved subgrade.

Penetration*: The consistency of a bituminous material expressed as the distance (in tenths of a millimeter) that a standard needle penetrates a sample vertically under specified conditions of loading, time and temperature.

Penetration Grading: A classification system of asphalt cements based on penetration in 0.1 mm at 25°C (77°F). There are five standard penetration grades for paving: 40-50, 60-70, 85-100, 120-150, and 200-300.

Performance Graded (PG): Asphalt binder grade designation used in Superpave; based on the binder's mechanical performance at critical temperatures and aging conditions. This system directly correlates laboratory testing to field performance through engineering principles.

Plant Screens*: Screens located between the dryer and hot bins which separate heated aggregates into proper hot bin sizes.

Poise: A centimeter-gram-second unit of absolute viscosity, equal to the viscosity of a fluid in which a stress of one dyne per square centimeter is required to maintain a difference of velocity of one centimeter per second between two parallel planes in the fluid that lie in the direction of flow and are separated by a distance of one centimeter.

Polymer Modified Asphalt Binder: A conventional asphalt cement to which a styrene block copolymer or styrene butadiene rubber (SBR) latex or neoprene latex has been added to improve performance.

Raveling: The progressive separation of aggregate particles in a pavement from the surface downward or from the edges inward. Raveling is caused by lack of compaction, construction of a thin lift during cold weather, dirty or disintegrating aggregate, too little asphalt in the mix, or overheating of the asphalt mix.

Ruts (Channels): Grooves that develop in the wheel tracks of a pavement. Channels may result from consolidation or lateral movement under traffic in one or more of the underlying courses, or by displacement in the asphalt surface layer itself. They may develop under traffic in new asphalt pavements that had too little compaction during construction or from plastic movement in a mix that does not have enough stability to support the traffic.

Sand Asphalt: A mixture of sand and asphalt cement, cutback asphalt or emulsified asphalt. It may be prepared with or without special control of aggregate grading and may or may not contain mineral filler. Either mixing-in-place or plant mix construction may be employed. Sand asphalt is used in construction of both base and surface courses.

Sheet Asphalt: A hot mixture of asphalt binder with clean, angular, graded sand and mineral filler. Its use is ordinarily confined to reservoir liners and land fill caps; usually laid on an intermediate or leveling course.

Sieve: In laboratory work, an apparatus in which the openings in the mesh are square for separating sizes of material.

Skid Resistance: The ability of an asphalt paving surface, particularly when wet, to offer resistance to slipping or skidding. The factors for obtaining high skid resistance are generally the same as those for obtaining high stability. Proper asphalt content and aggregate with a rough surface texture are the greatest contributors. The aggregate must not only have a rough surface texture, but also resist polishing.

Solubility: A measure of the purity of an asphalt cement. The ability of the portion of the asphalt cement that is soluble to be dissolved in a specified solvent.

Source Properties: Critical aggregate characteristics, which because of their nature, are source specific and whose use and limiting values are source dependent and established by the using agency.

Stability: The ability of an asphalt paving mixtures to resist deformation from imposed loads. Stability is dependent upon both internal friction and cohesion.

Stoke: A unit of kinematic viscosity, equal to the viscosity of a fluid in poises divided by the density of the fluid in grams per cubic centimeter.

Subbase: The course in the asphalt pavement structure immediately below the base course. If the subgrade soil has adequate support, it may serve as the subbase.

Subgrade: The soil prepared to support a pavement structure or a pavement system. It is the foundation of the pavement structure.

Subgrade, Improved: Subgrade that has been improved as a working platform by: 1) the incorporation of granular materials or stabilizers such as asphalt, lime, or portland cement, into the subgrade soil; or 2) any course or courses of select or improved material placed on the subgrade soil below the pavement structure. Subgrade improvement does not affect the design thickness of the pavement structure.

Superpave™: Short for "Superior Performing Asphalt Pavement" – a performance-based system for selecting and specifying asphalt binders and for developing an asphalt mixture design.

Superpave Gyratory Compactor (SGC): A device used during Superpave mix design or field testing activities for compacting samples of hot mix asphalt into specimens used for volumetric analysis. Continuous densification of the specimen is measured during the compaction process.

Superpave Mix Design: A mixture design system that integrates the selection of materials (asphalt, aggregate) and volumetric proportioning with the project's climate and design traffic.

Viscosity: A measure of the resistance to flow of a liquid. It is one method of measuring the consistency of asphalt.

> **Absolute Viscosity:** A measure of the viscosity of asphalt, measured in poises, conducted at a temperature of 60°C (140°F). The test method utilizes a partial vacuum to induce flow in the viscometer.

> **Kinematic Viscosity:** A measure of the viscosity of asphalt, measured in centistokes, conducted at a temperature of 135°C (275°F).

Viscosity Grading: A classification system of asphalt cements based on viscosity ranges at 60°C (140°F). A minimum viscosity at 135°C (275°F) is also usually specified. The purpose is to prescribe limiting values of consistency at these two temperatures. 60°C (140°F) approximates the maximum temperature of an asphalt pavement surface in service in the U.S.; 135°C (275°F) approximates the mixing and laydown temperatures for hot mix asphalt pavements.

Wet Mixing Period: The interval of time between the beginning of application of asphalt material into a pugmill and the opening of the discharge gate.

Workability: The ease with which paving mixtures may be placed and compacted.

See examples in Chapter 7
for discussion of using random
number tables.

APPENDIX C

RANDOM NUMBER TABLES

Table C.01 Random Number Table

	Col No. 1			Col No. 2			Col No. 3			Col No. 4			Col No. 5			Col No. 6			Col No. 7	
A	B	C	A	B	C	A	B	C	A	B	C	A	B	C	A	B	C	A	B	C
15	.033	.576	05	.048	.879	21	.013	.220	18	.089	.716	17	.024	.863	30	.030	.901	12	.029	.386
21	.101	.300	17	.074	.156	30	.036	.853	10	.102	.330	24	.060	.032	21	.096	.198	18	.112	.284
23	.129	.916	18	.102	.191	10	.052	.746	14	.111	.925	26	.074	.639	10	.100	.161	20	.114	.848
30	.158	.434	06	.105	.257	25	.061	.954	28	.127	.840	07	.167	.512	29	.133	.388	03	.121	.656
24	.177	.397	28	.179	.447	29	.062	.507	24	.132	.271	28	.194	.776	24	.138	.062	13	.178	.640
11	.202	.271	26	.187	.844	18	.087	.887	19	.285	.899	03	.219	.166	20	.168	.564	22	.209	.421
16	.204	.012	04	.188	.482	24	.105	.849	01	.326	.037	29	.264	.284	22	.232	.953	16	.221	.311
08	.208	.418	02	.208	.577	07	.139	.159	30	.334	.938	11	.282	.262	14	.259	.217	29	.235	.356
19	.211	.798	03	.214	.402	01	.175	.641	22	.405	.295	14	.379	.994	01	.275	.195	28	.264	.941
29	.233	.070	07	.245	.08	23	.196	.873	05	.421	.282	13	.394	.405	06	.277	.475	11	.287	.199
07	.260	.073	15	.248	.831	26	.24	.981	13	.451	.212	06	.410	.157	02	.296	.497	02	.336	.992
17	.262	.308	29	.261	.087	14	.255	.374	02	.461	.023	15	.438	.70	26	.311	.144	15	.393	.488
25	.271	.180	30	.302	.883	06	.31	.043	06	.487	.539	22	.453	.635	05	.351	.141	19	.437	.655
06	.302	.372	21	.318	.088	11	.316	.653	08	.497	.396	21	.472	.824	17	.370	.811	24	.466	.773
01	.409	.406	11	.376	.936	13	.324	.585	25	.503	.893	05	.488	.118	09	.388	.484	14	.531	.014
13	.507	.693	14	.43	.814	12	.351	.275	15	.594	.603	01	.525	.222	04	.410	.073	09	.562	.678
02	.575	.654	27	.438	.676	20	.371	.535	27	.620	.894	12	.561	.980	25	.471	.530	06	.601	.750
18	.591	.318	08	.467	.205	08	.409	.495	21	.629	.841	08	.652	.508	13	.486	.779	10	.612	.859
20	.610	.821	09	.474	.138	16	.445	.740	17	.691	.583	18	.668	.271	15	.515	.867	26	.673	.112
12	.631	.597	10	.492	.474	03	.474	.929	09	.708	.689	30	.736	.634	23	.567	.798	23	.738	.770
27	.651	.281	13	.499	.892	27	.543	.387	07	.709	.012	02	.763	.253	11	.618	.502	21	.753	.614
04	.661	.953	19	.511	.520	17	.625	.171	11	.714	.049	23	.804	.140	28	.636	.148	30	.758	.851
22	.692	.089	23	.591	.770	02	.699	.073	23	.720	.695	25	.828	.425	27	.65	.741	27	.765	.563
05	.779	.346	20	.604	.730	19	.702	.934	03	.748	.413	10	.843	.627	16	.711	.508	07	.780	.534
09	.787	.173	24	.654	.330	22	.816	.802	20	.781	.603	16	.858	.849	19	.778	.812	04	.818	.187
10	.818	.837	12	.728	.523	04	.838	.166	26	.830	.384	04	.903	.327	07	.804	.675	17	.837	.353
14	.895	.631	16	.753	.344	15	.904	.116	04	.843	.002	09	.912	.382	08	.806	.952	05	.854	.818
26	.912	.376	01	.806	.134	28	.969	.740	12	.884	.582	27	.935	.162	18	.841	.414	01	.867	.133
28	.920	.163	22	.878	.884	09	.974	.046	29	.926	.700	20	.970	.582	12	.918	.114	08	.915	.538
03	.945	.14	25	.939	.162	05	.977	.494	16	.951	.601	19	.975	.327	03	.992	.399	25	.975	.584

	Col No. 8			Col No. 9			Col No. 10			Col No. 11			Col No. 12			Col No. 13			Col No. 14	
A	B	C	A	B	C	A	B	C	A	B	C	A	B	C	A	B	C	A	B	C
09	.042	.071	14	.061	.935	26	.038	.023	27	.074	.779	16	.073	.987	03	.033	.091	26	.035	.175
17	.141	.411	02	.065	.097	30	.066	.371	06	.084	.396	23	.078	.056	07	.047	.391	17	.089	.363
02	.143	.221	03	.094	.228	27	.073	.876	24	.098	.524	17	.096	.076	28	.064	.113	10	.149	.681
05	.162	.899	16	.122	.945	09	.095	.568	10	.133	.919	04	.153	.163	12	.066	.360	28	.238	.075
03	.285	.016	18	.158	.43	05	.180	.741	15	.187	.079	10	.254	.834	26	.076	.552	13	.244	.767
28	.291	.034	25	.193	.469	12	.200	.851	17	.227	.767	06	.284	.628	30	.087	.101	24	.262	.366
08	.369	.557	24	.224	.572	13	.259	.327	20	.236	.571	12	.305	.616	02	.127	.187	08	.264	.651
01	.436	.386	10	.225	.223	21	.264	.681	01	.245	.988	25	.319	.901	06	.144	.068	18	.285	.311
20	.450	.289	09	.233	.838	17	.283	.645	04	.317	.291	01	.320	.212	25	.202	.674	02	.340	.131
18	.455	.789	20	.29	.120	23	.363	.063	29	.350	.911	08	.416	.372	01	.247	.025	29	.353	.478
23	.488	.715	01	.297	.242	20	.364	.366	26	.380	.104	13	.432	.556	23	.253	.323	06	.359	.270
14	.496	.276	11	.337	.760	16	.395	.363	28	.425	.864	02	.489	.827	24	.320	.651	20	.387	.248
15	.503	.342	19	.389	.064	02	.423	.540	22	.487	.526	29	.503	.787	10	.328	.365	14	.392	.694
04	.515	.693	13	.411	.474	08	.432	.736	05	.552	.511	15	.518	.717	27	.338	.412	03	.408	.770
16	.532	.112	20	.447	.893	10	.476	.468	14	.564	.357	28	.524	.998	13	.356	.991	27	.440	280
22	.557	.357	22	.478	.321	03	.508	.774	11	.572	.306	03	.542	.362	16	.401	.792	22	.461	.830
11	.559	.620	29	.481	.993	01	.601	.417	21	.594	.197	19	.585	.462	17	.423	.117	16	.527	.003
12	.650	.216	27	.562	.403	22	.687	.917	09	.607	.524	05	.695	.111	21	.481	.838	30	.531	.486
21	.672	.320	04	.566	.179	29	.697	.862	19	.650	.572	07	.733	.838	08	.560	.401	25	.678	.360
13	.709	.273	08	.603	.758	11	.701	.605	18	.664	.101	11	.744	.948	19	.564	.190	21	.725	.014
07	.745	.687	15	.632	.927	07	.728	.498	25	.674	.428	18	.793	.748	05	.571	.054	05	.797	.595
30	.780	.285	06	.707	.107	14	.745	.679	02	.697	.674	27	.802	.967	18	.587	.584	15	.801	.927
19	.845	.097	28	.737	.161	24	.819	.444	03	.767	.928	21	.826	.487	15	.604	.145	12	.836	.294
26	.846	.366	17	.846	.13	15	.84	.826	16	.809	.529	24	.835	.832	11	.641	.298	04	.854	.982
29	.861	.307	07	.874	.491	25	.863	.568	030	.838	.294	26	.855	.142	22	.672	.156	11	.884	.928
25	.906	.874	05	.880	.828	06	.878	.215	013	.845	.470	14	.861	.462	20	.674	.887	19	.886	.832
24	.919	.809	23	.931	.659	18	.93	.601	08	.855	.524	20	.874	.625	14	.752	.881	07	.929	.932
10	.952	.555	26	.960	.365	04	.954	.827	07	.867	.718	30	.929	.056	09	.774	.560	09	.932	.206
06	.961	.504	21	.978	.194	28	.963	.004	12	.881	.722	09	.935	.582	29	.921	.752	01	.970	.692
27	.969	.811	12	.982	.183	19	.988	.02	23	.937	.872	22	.947	.797	04	.959	.099	23	.973	.082

Table C.01 Random Number Table (Continued)

Col No. 15 A	B	C	Col No. 16 A	B	C	Col No. 17 A	B	C	Col No. 18 A	B	C	Col No. 19 A	B	C	Col No. 20 A	B	C	Col No. 21 A	B	C
15	.023	.979	19	.062	.588	13	.045	.004	25	.027	.290	12	.052	.075	20	.030	.881	01	.010	.946
11	.118	.465	25	.080	.218	18	.086	.878	06	.057	.571	30	.075	.493	12	.034	.291	10	.014	.939
07	.134	.172	09	.131	.295	26	.126	.990	26	.059	.026	28	.120	.341	22	.043	.893	09	.032	.346
01	.139	.230	18	.136	.381	12	.136	.661	07	.105	.176	27	.145	.689	28	.143	.073	06	.093	.180
16	.145	.122	05	.147	.864	30	.146	.337	18	.107	.358	02	.209	.957	03	.150	.937	15	.151	.012
20	.165	.520	12	.158	.365	05	.169	.470	22	.128	.827	26	.272	.818	04	.154	.867	16	.185	.455
06	.185	.481	28	.214	.184	21	.244	.433	23	.156	.440	22	.299	.317	19	.158	.359	07	.227	.277
09	.211	.316	14	.215	.757	23	.270	.849	15	.171	.157	18	.306	.475	29	.304	.615	02	.304	.400
14	.248	.348	13	.224	.846	25	.274	.407	08	.220	.097	20	.311	.653	06	.369	.633	30	.316	.074
25	.219	.890	15	.227	.809	10	.290	.925	20	.252	.066	15	.348	.156	18	.390	.536	18	.328	.799
13	.252	.577	11	.280	.898	01	.323	.490	04	.268	.576	16	.381	.710	17	.403	.392	20	.352	.288
30	.273	.088	01	.331	.925	24	.352	.291	14	.275	.302	01	.411	.607	23	.404	.182	26	.371	.216
18	.277	.689	10	.399	.992	15	.361	.155	11	.297	.589	13	.417	.715	01	.415	.457	19	.448	.754
22	.372	.958	30	.417	.787	29	.374	.882	01	.358	.305	21	.472	.484	07	.437	.696	13	.487	.598
10	.461	.075	08	.439	.921	08	.432	.139	09	.412	.089	4	.478	.885	24	.446	.546	12	.546	.640
28	.519	.536	20	.472	.484	04	.467	.266	16	.429	.834	25	.479	.080	26	.485	.768	24	.550	.038
17	.520	.090	24	.498	.712	22	.508	.880	10	.491	.203	11	.566	.104	15	.511	.313	03	.604	.780
03	.523	.519	04	.516	.396	27	.632	.191	28	.542	.306	10	.576	.659	10	.517	.290	22	.621	.930
26	.573	.502	03	.548	.688	16	.661	.836	12	.563	.091	29	.665	.397	30	.556	.853	21	.629	.154
19	.634	.206	23	.597	.508	19	.675	.629	02	.593	.321	19	.739	.298	25	.561	.837	11	.634	.908
24	.635	.810	21	.681	.114	14	.680	.890	30	.692	.198	14	.749	.759	09	.574	.599	05	.696	.459
21	.679	.841	02	.739	.298	28	.714	.508	19	.705	.445	08	.756	.919	13	.613	.762	23	.710	.078
27	.712	.366	29	.792	.038	06	.719	.441	24	.709	.717	07	.798	.183	11	.698	.783	29	.726	.585
05	.780	.497	22	.829	.324	09	.735	.040	13	.820	.739	23	.834	.647	14	.715	.179	17	.749	.916
23	.861	.106	17	.834	.647	17	.741	.906	05	.848	.866	06	.837	.978	16	.770	.128	04	.802	.186
12	.865	.377	16	.909	.608	11	.747	.205	27	.867	.633	03	.849	.964	08	.815	.385	14	.835	.319
29	.882	.635	06	.914	.420	20	.850	.047	03	.883	.333	24	.851	.109	05	.872	.490	08	.870	.546
08	.902	.020	27	.958	.856	02	.859	.356	17	.900	.443	05	.859	.935	21	.885	.999	28	.871	.539
04	.951	.482	26	.981	.976	07	.870	.612	21	.914	.483	17	.863	.220	02	.958	.177	25	.971	.369
02	.977	.172	07	.983	.624	03	.916	.463	29	.950	.753	09	.863	.147	27	.961	.980	27	.984	.252

Col No. 22 A	B	C	Col No. 23 A	B	C	Col No. 24 A	B	C	Col No. 25 A	B	C	Col No. 26 A	B	C	Col No. 27 A	B	C	Col No. 28 A	B	C
12	.051	.032	26	.051	.187	08	.015	.521	02	.039	.005	16	.026	.102	21	.050	.952	29	.042	.039
11	.068	.980	03	.053	.256	16	.068	.994	16	.061	.599	01	.033	.886	17	.085	.403	07	.105	.293
17	.089	.309	29	.100	.159	11	.118	.400	26	.068	.054	04	.088	.686	10	.141	.624	25	.115	.420
01	.091	.371	13	.102	.465	21	.124	.565	11	.073	.812	22	.090	.602	05	.154	.157	09	.126	.612
10	.100	.709	24	.110	.316	18	.153	.158	07	.123	.649	13	.114	.614	06	.164	.841	10	.205	.144
30	.121	.744	18	.114	.300	17	.190	.159	05	.126	.658	20	.136	.576	07	.197	.013	03	.210	.054
02	.166	.056	11	.123	.208	26	.192	.676	14	.161	.189	05	.138	.228	16	.215	.363	23	.234	.533
23	.179	.529	09	.138	.182	01	.237	.030	18	.166	.040	10	.216	.565	08	.222	.520	13	.266	.799
21	.187	.051	06	.194	.115	12	.283	.077	28	.218	.171	02	.233	.610	13	.269	.477	20	.305	.603
22	.205	.543	22	.234	.480	03	.286	.318	06	.255	.117	07	.278	.357	02	.288	.012	05	.372	.223
28	.230	.688	20	.274	.107	10	.317	.734	15	.261	.928	30	.405	.273	25	.333	.633	26	.385	.111
19	.243	.001	21	.331	.292	05	.337	.844	10	.301	.811	06	.421	.807	28	.348	.710	30	.422	.315
27	.267	.990	08	.346	.085	25	.441	.336	24	.363	.025	12	.426	.583	20	.362	.961	17	.453	.783
15	.283	.440	27	.382	.979	27	.469	.786	22	.378	.792	08	.471	.708	14	.511	.989	02	.460	.916
16	.352	.089	07	.387	.865	24	.473	.237	27	.379	.959	18	.473	.738	26	.540	.903	27	.461	.841
03	.377	.648	28	.411	.776	20	.475	.761	19	.420	.557	19	.510	.207	27	.587	.643	14	.483	.095
06	.397	.769	16	.444	.999	06	.557	.001	21	.467	.943	03	.512	.329	12	.603	.745	12	.507	.375
09	.409	.428	04	.515	.993	07	.610	.238	17	.494	.225	15	.640	.329	29	.619	.895	28	.509	.748
14	.465	.406	17	.518	.827	09	.617	.041	09	.620	.081	09	.665	.354	23	.623	.333	21	.583	.804
13	.499	.651	05	.539	.620	13	.641	.648	30	.623	.106	14	.680	.884	22	.624	.076	22	.587	.993
04	.539	.972	02	.623	.271	22	.664	.291	03	.623	.777	26	.703	.622	18	.670	.904	16	.689	.339
18	.560	.747	30	.637	.374	04	.668	.856	08	.651	.790	29	.739	.394	11	.711	.253	06	.727	.298
26	.575	.892	14	.714	.364	19	.717	.232	12	.715	.599	25	.759	.386	01	.790	.392	04	.731	.814
29	.756	.712	15	.730	.107	02	.776	.504	23	.782	.093	24	.803	.602	04	.813	.611	08	.807	.983
20	.760	.920	19	.771	.552	29	.777	.548	20	.810	.371	27	.842	.491	19	.843	.732	15	.833	.757
05	.847	.925	23	.780	.662	14	.823	.223	01	.841	.726	21	.870	.435	03	.844	.511	19	.896	.464
25	.872	.891	10	.924	.888	23	.848	.264	29	.862	.009	28	.906	.367	30	.858	.299	18	.916	.384
24	.874	.135	12	.929	.204	30	.892	.817	25	.891	.873	23	.948	.367	09	.929	.199	01	.948	.610
08	.911	.215	01	.937	.714	28	.943	.19	04	.917	.264	11	.956	.142	24	.931	.263	11	.976	.799
07	.946	.065	25	.974	.398	15	.975	.962	13	.958	.990	17	.993	.989	15	.939	.947	24	.978	.633

Table D.01 AASHTO and ASTM Test Methods As Commonly Specified

	AASHTO	ASTM
ASPHALTS		
Asphalt Binder/Cement		
Performance Graded Asphalt Binder	MP 1	——
Viscosity, Kinematic	T 201	D 2170
Viscosity, Absolute	T 202	D 2171
Penetration	T 49	D 5
Flash Point, Cleveland Open Cup	T 48	D 92
Flash Point, Pensky-Martens	T 73	D 93
Solubility	T 44	D 2042
Thin Film Oven	T 179	D 1754
Rolling Thin Film Oven	T 240	D 2872
Ductility	T 51	D 113
Specific Gravity	T 228	D 70
Softening Point	T 53	D 2398
Cutback Asphalt		
Viscosity, Kinematic at 60°C	T 201	D 2170
Flash Point, Tag Open Cup	T 79	D 1310
Distillation	T 78	D 402
Tests on Residue:		
Viscosity, Absolute	T 202	D 2171
Ductility	T 51	D 113
Solubility	T 44	D 2042
Water in Asphalt	T 55	D 95
Specific Gravity	T 227	D 1298
Emulsified Asphalt		
Viscosity, Saybolt Furol	T 59	D 244
Storage Stability	T 59	D 244
Demulsibility	T 59	D 244
Coating Ability and Water Resistance	T 59	D 244
Particle Charge Test	T 59	D 244
Sieve Test	T 59	D 244
Residue by Distillation	T 59	D 244
Oil Distillate	T 59	D 244
Tests on Residue:		
Penetration	T 59	D 244
Ductility	T 59	D 244
Solubility	T 59	D 244
Float Test	T 59	D 244
MINERAL AGGREGATES		
Sieve Analysis of Fine and Coarse Aggregates	T 27	C 136
Sieve Analysis of Mineral Filler	T 37	D 546
Sand Equivalent	T 176	D 2419
Resistance to Abrasion (Los Angeles Machine)	T 96	C 131
Soundness (Sodium Sulfate or Magnesium)	T 104	C 88

Table D.01 AASHTO and ASTM Test Methods as Commonly Specified (Continued)

	AASHTO	ASTM
Specific Gravity:		
Coarse Aggregate	T 85	C 127
Fine Aggregate	T 84	C 128
Filler	T 100 or	D 854 or
	T 133	C 188
Unit Weight	T 19	C 29
Moisture Content	T 255	C 566

HOT MIX ASPHALT

	AASHTO	ASTM
Superpave Volumetric Mix Design	MP 2	——
Marshall Mix Design:		
Resistance to Plastic Flow (Stability and Flow)	T 245	D 1559
Bulk Specific Gravity of Compacted Specimens	T 166	D 2726
Percent Air Voids	T 269	D 3203
Maximum Specific Gravity	T 209	D 2041
Extraction of Bitumen	T 164	D 2172
Recovery of Asphalt (Abson Method)	T 170	D 1856
Moisture or Volatile Distillates	T 110	D 1461
Hveem Mix Design:		
Preparation of Test Specimens with Kneading Compactor	T 247	D 1561
Resistance to Deformation and Cohesion	T 246	D 1560

APPENDIX E

SUPERPAVE PERFORMANCE GRADED ASPHALT BINDER SPECIFICATIONS

Performance Grade	PG 46			PG 52							PG 58					PG 64						PG 70						PG 76					PG 82				
	-34	-40	-46	-10	-16	-22	-28	-34	-40	-46	-16	-22	-28	-34	-40	-10	-16	-22	-28	-34	-40	-10	-16	-22	-28	-34	-40	-10	-16	-22	-28	-34	-10	-16	-22	-28	-34
Average 7-day Maximum Pavement Design Temperature, °C [a]	<46			<52							<58					<64						<70						<76					<82				
Minimum Pavement Design Temperature, °C [a]	>-34	>-40	>-46	>-10	>-16	>-22	>-28	>-34	>-40	>-46	>-16	>-22	>-28	>-34	>-40	>-10	>-16	>-22	>-28	>-34	>-40	>-10	>-16	>-22	>-28	>-34	>-40	>-10	>-16	>-22	>-28	>-34	>-10	>-16	>-22	>-28	>-34
Original Binder																																					
Flash Point Temp, T 48: Minimum °C	230																																				
Viscosity, ASTM D 4402: Maximum, 3 Pa·s (3000 cP), Test Temp, °C	135																																				
Dynamic Shear, TP 5: [c] G*/sin δ, Minimum, 1.00 kPa Test Temperature @ 10 rad/s, °C	46			52							58					64						70						76					82				
Rolling Thin Film Oven (T 240) or Thin Film Oven (T 179) Residue																																					
Mass Loss, Maximum, %	1.00																																				
Dynamic Shear, TP 5: G*/sin δ, Minimum, 2.20 kPa Test Temperature @ 10 rad/s, °C	46			52							58					64						70						76					82				
Pressure Aging Vessel Residue (PP 1)																																					
PAV Aging Temperature, °C [d]	90			90							100					100						100 (110)						100 (110)					100 (110)				
Dynamic Shear, TP 5: G*/sin δ, Maximum, 5000 kPa Test Temperature @ 10 rad/s, °C	10	7	4	25	22	19	16	13	10	7	25	22	19	16	13	31	28	25	22	19	16	34	31	28	25	22	19	37	34	31	28	25	40	37	34	31	28
Physical Hardening [e]	Report																																				
Creep Stiffness, TP1: [f] S, Maximum, 300 MPa m-value, Minimum, 0.300 Test Temperature @ 60 sec, °C	-24	-30	-36	0	-6	-12	-18	-24	-30	-36	-6	-12	-18	-24	-30	0	-6	-12	-18	-24	-30	0	-6	-12	-18	-24	-30	0	-6	-12	-18	-24	0	-6	-12	-18	-24
Direct Tension, TP 3 [f] Failure Strain, Minimum, 1.0% Test Temp @ 1.0 mm/min, °C	-24	-30	-36	-0	-6	-12	-18	-24	-30	-36	-6	-12	-18	-24	-30	0	-6	-12	-18	-24	-30	0	-6	-12	-18	-24	-30	0	-6	-12	-18	-24	0	-6	-12	-18	-24

Notes:

a. Pavement temperatures can be estimated from air temperatures using an algorithm contained in the Superpave™ software program or may be provided by the specifying agency, or by following the procedures as outlined in MP 2 and PP 28.

b. This requirement may be waived at the discretion of the specifying agency if the supplier warrants that the asphalt binder can be adequately pumped and mixed at temperatures that meet all applicable safety standards.

c. For quality control of unmodified asphalt cement production, measurement of the viscosity of the original asphalt cement may be substituted for dynamic shear measurements of G*/sin δ at test temperatures where the asphalt is a Newtonian fluid. Any suitable standard means of viscosity measurement may be used, including capillary or rotational viscometer (AASHTO T 201 or T 202).

d. The PAV aging temperature is based on simulated climatic conditions and is one of three temperatures 90°C, 100°C or 110°C. The PAV aging temperature is 100°C for PG 64- and above, except in desert climates, where it is 110°C.

e. Physical Hardening - TP 1 is performed on a set of asphalt beams according to Section 13.1 of TP 1, except the conditioning time is extended to 24 hrs ± 10 minutes at 10°C above the minimum performance temperature. The 24-hour stiffness and m-value are reported for information purposes only.

f. If the creep stiffness is below 300 MPa, the direct tension test is not required. If the creep stiffness is between 300 and 600 MPa, the direct tension failure strain requirement can be used in lieu of the creep stiffness requirement. The m-value requirement must be satisfied in both cases.

From AASHTO Provisional Standards, Third Edition, Copyright 1996, by the American Association of State Highway and Transportation Officials, Washington, D.C. Used by permission.

CPSIA information can be obtained
at www.ICGtesting.com
Printed in the USA
FFHW012300241118
49449852-53810FF